OXFORD RESEARCH STUDIES IN GEOGRAPHY

General Editors

J. Gottmann	F. V. Emery
J. A. Steers	C. D. Harris

Sea-Level Changes

North-West England
during the Flandrian Stage

M. J. Tooley

CLARENDON PRESS · OXFORD
1978

Oxford University Press, Walton Street, Oxford OX2 6DP

OXFORD LONDON GLASGOW
NEW YORK TORONTO MELBOURNE WELLINGTON
IBADAN NAIROBI DAR ES SALAAM LUSAKA CAPE TOWN
KUALA LUMPUR SINGAPORE JAKARTA HONG KONG TOKYO
DELHI BOMBAY CALCUTTA MADRAS KARACHI

© *Oxford University Press 1978*

British Library Cataloguing in Publication Data
Tooley, M. J.
    Sea-level changes. — (Oxford research studies in geography).
    1. Sea Level — Irish Sea   2. Sea level — England
    1. Title   II. Series
    551.4'61'37   GC591   77-30248

ISBN 0-19-823228-4

Set by Malvern Typesetting Services Ltd
and printed in Great Britain by
Cox & Wyman Ltd,
London, Fakenham and Reading

for
ROSANNA MARY TOOLEY
and
NICHOLAS WILLIAM TOOLEY

EDITORIAL PREFACE

Since the end of World War II there has been great interest taken in the study of our coastline. There are many practical applications of this work. Not until the results of the great flood of 1953 were analysed and made known to inland dwellers were the seriousness of erosion by the sea, and the oscillations to which sea-level is subject, realised by fewer than one per cent of the population. The recent floods (November 1977 and January 1978) on the east and west coasts have once again emphasised the importance of what may happen on the coast.

Changes of sea-level relative to the land have been in continuous action throughout geological time. In the present context we can ignore those which took place before the Tertiary or even later, not because they were unimportant but because it is those which have left their imprint on the coast during the Quaternary that give us so much information about recent changes of level of land and sea. The Waverly Committee in its final report (1954) recommended 'that, in regard to the Thames, investigations should be undertaken urgently into the possibility of providing a suitable structure, capable of being closed, as a means of reducing the maximum water levels higher up the river at the time of a surge.' The earlier (1928) severe flooding in London led to a great deal of discussion but little was done to anticipate future trouble. It is common knowledge that much of Roman London is now below high water level, and that the walls along the lower river have been raised from time to time. The south east of England is slowly down-tilted whilst in the north and in Scotland there is evidence of a reverse movement.

This difference between north and south emphasises the importance of Dr Tooley's work on the north-western coasts of the country. Many investigators in the past and today are particularly interested in unravelling the field evidence of those coastal oscillations, and Sir Harry Godwin called attention to the significance of estuaries and shallow inlets in this type of work. The coast between North Wales and St. Bees Head is admirably suited for such a study. Dr Tooley's knowledge of this stretch is profound, and he has made most careful and detailed investigations of 36 sites. Work of this nature involves not only an appreciation of several techniques — pollen analysis and the methods of dating deposits, etc. — but also an understanding of the physical, geological and botanical nature of the deposits left by the sea in its transgressive and regressive phases. In this book he has followed Troels-Smith's notation which allows an adequate 'shorthand' description in a small space. For reasons of economy, not all of the author's tables have been reproduced.

Because Morecambe Bay is on or close to a hinge-line — to the north the coast has risen; to the south it is sinking — the information that Dr Tooley has gathered is remarkably interesting in itself and of obvious significance in the general problem of vertical coastal movements in these islands and in north-western Europe. Moreover, the most complete sequence of marine and nearly marine sediments for the Flandrian stage is found in the lagoonal and tidal flat zones of Lytham, beneath an overburden of dune sands or beneath recently reclaimed estuarine marshes. The author makes brief comparisons with the Fenland, Somerset Levels and some continental localities.

This volume is a work of scholarship; it is concisely and clearly written and demands careful study since a proper understanding of its contents may well help us not only to appreciate what has happened, but also to enable us, at any rate to some extent, to anticipate what may happen in the future. It cannot be over-emphasised that an adequate understanding of eustatic and isostatic movements of sea-level considered in relation to tidal, meteorological and surge phenomena, and the relation of all to river floods and to the rate of erosion along many miles of low coasts or coasts fronted by easily eroded cliffs, make this a study of national and not just of local importance.

March 1978   J.G.   J.A.S.
                                                    F.V.E.   C.D.H.

PREFACE

In 1974, the Board of the International Geological Correlation Programme (I.G.C.P.) approved a key project on *Sea-level movements during the last 15,000 years*, and thereby initiated a programme of co-ordinated research into global sea-level movements (Bloom, 1974; Tooley, 1976a).

The subject of this research monograph is a study of the evidence for, and nature of, sea-level changes in north-west England. The methodology that has been developed and applied to this problem may be of value in scientific investigations on sea-level changes elsewhere in the world, and the data contribute to the information available for correlation in the I.G.C.P. project.

In north-west England and North Wales, low-lying coastal areas and extensive tidal flats have attracted examination and observation for the last two hundred years: a summary of some of the more apposite results is given in Chapter 1. The techniques that were found to be most useful in elucidating problems associated with sea-level studies were plane surveying, stratigraphic, pollen and diatom analyses and radiocarbon dating: these techniques and the sources of error to which the raw data were subject are also considered in Chapter 1. A selection of the more important and critical sites in west Lancashire is presented in Chapter 2, in order to demonstrate how the data were derived for the index points on the sea-level curve and the establishment of coastal sequences. In Chapter 3, the age and extent of marine transgressions in west Lancashire are described as a basis for correlation throughout north-west England. The latitudinal spread of data allows a comparison to be made of the altitudinal variations of the transgression surfaces, not only within north-west England, but also in south-west Scotland, Wales and the Severn estuary.

Although evidence for sea-level changes has been collected from an extensive area in north-west England and North Wales, the index points for the sea-level curve described in Chapter 4 are located in a restricted geographical area flanking the estuary of the River Ribble. That the area is confined, fulfils one of the requirements of the *Sea-level movements* project that sea-level studies should be within small homogeneous areas.

In Chapter 4 also, some consideration is given to the nature of the *Hillhouse coastline* in the light of the evidence from the stratigraphy, pollen, and dating evidence given in Chapter 2. A scheme for the evolution of the Lancashire coast is proposed and this may serve as a framework for future work on sea-level movements and transgression sequences in this area.

When coastal sequences and a sea-level curve have been established in a restricted area, their relationships to patterns in adjacent areas may be considered. In Chapter 5, correlations are suggested with well-documented sites in north-east England, the Fenlands and Somerset Levels in Britain, and in France, the Netherlands and Sweden.

Finally, the relationship between sea-level movements and climatic change is explored, and tentative correlations established between the sea-level curve from west Lancashire and additional marine and terrestrial evidence for climatic change.

ACKNOWLEDGEMENTS

I gained an interest and love of the coast and of the sea in North Devon. It was only a short distance from Barnstaple to Croyde, Baggy Point, Saunton Down, Saunton Sands and Braunton Burrows, where we spent most week-ends. Here the interglacial raised beach was exposed in cliff sections and passed beneath the dunes at Croyde and Saunton. In 1956, we moved to the Fylde coast, in Lancashire, and it was here that I began to appreciate tidal and sea-level changes, and the mosaic of distinctive landscapes within the township of Lytham—the sand-dunes, saltmarshes, and reclaimed peat mosses. My parents' house in St. Annes was at a point, some six metres above Ordnance Datum, where sand had overblown the moss, and by excavation it was possible to reach and sample the peat and clay beneath, establishing the plant succession following the end of a marine transgression. This site subsequently became Lytham Common—1.

To my father, W. A. Tooley, a chartered engineer, I am indebted for teaching me the rudiments of surveying, assisting in the collection of field data, and stressing the need for precise and systematic field recording; and to my mother for her unstinted encouragement and support.

With this background, I began work on sea-level changes in north-west England, whilst I was in receipt of an N.E.R.C. studentship, held in the Department of Environmental Sciences at the University of Lancaster from 1966 until 1969. It is a pleasure to acknowledge the guidance and assistance here and subsequently of Professor G. Manley, Professor F. Oldfield and Dr. Ada W. Pringle (née Phillips). In 1968, I participated in the British Council Exchange scheme, and visited the Geological Survey of the Netherlands, where I was taught diatom analysis by Mr. A. du Saar and shown critical coastal sections by Dr. Saskia Jelgersma and Mr. J. F. Van Regteren Altena. This visit and the return visit of Dr. Jelgersma to Lancashire, confirmed me in the procedures of field and laboratory analysis, that were necessary to elucidate sea-level changes. Through Dr. Jelgersma's recommendations, radiocarbon dating facilities were obtained with Dr. M. A. Geyh at *Niedersächsisches Landesamt für Bodenforschung* in Hanover, and without these facilities much of the scale and scope of the present work would have been impossible. To both Dr. Geyh and Dr. Jelgersma, I am grateful.

From 1969 until 1972, when the first manuscript of this book had been completed, additional field and laboratory work was carried out in the Department of Geography, University of Durham, and I am grateful to the head of Department, Professor W. B. Fisher, for the research and technical facilities made available to me: Mr. J. Telford assisted in the field and carried out preparations for pollen analysis, Mr. G. Brown drew most of the figures, Mr. D. Hudspeth photographed and reduced the figures, and Mrs. S. Eckford and Miss B. Egginton typed the manuscript. A grant from the Staff Research Fund of Durham University enabled me to obtain deep cores from the Fylde coast. My wife, Rosanna, assisted by correcting the final manuscript and creating the illustration for the jacket.

Acknowledgement is made to the Royal Geographical Society to reproduce Figs. 5, 6, 33 and 34, and to the Editor of the *Geological Journal* to reproduce Figs. 11, 12, 13 and 37.

It is a pleasure to record the criticism, advice, assistance, and encouragement of many colleagues and friends at various times during the eight years that this work was in preparation: Dr. J. T. Andrews, Dr. B. Berglund, Mr. and Mrs. J. Cherry, Dr. D. E. Cotton, Dr. and Mrs. P. Cundill, Dr. S. Funder, the late Miss M. Garnett, Dr. D. Gilbertson, Mr. A. R. Gunson, Dr. D. Huddart, Dr. B. S. John, the late Mrs. E. M. Megaw, Dr. N.-A. Mörner, Mr. J. Pogson, Dr. D. E. Smith, Dr. M. Ters, Dr. J. Troels-Smith and Dr. R. G. West F.R.S. I have also been fortunate in the students I have taught at Durham, some of whom have carried out work on coastal problems: in particular, those who have assisted in the collection of field data include, Mr. D. Chester, Dr. R. J. N. Devoy, Miss P. A. GreatRex, the late Mr. S. Hogg, Mr. J. Ince, Mr. D. M. Paterson and Mr. J. Pelican.

All Quaternary scientists who have worked in the British Isles owe no small debt to the work of Professor Sir Harry Godwin, F.R.S. Likewise, this contribution has been stimulated by Sir Harry's research on sea-level changes in the Fenlands and Somerset Levels. I am also grateful to him for his criticisms, observations and encouragement as this work has progressed.

Finally, I must record my gratitude to Professor J. A. Steers, who has made many helpful observations, and has guided this work to publication.

M. J. TOOLEY

Witton-le-Wear
October 1976

In March 1977, a symposium on The Quaternary history of the Baltic, the North Sea and the Irish Sea was held in Sweden to celebrate the 500th anniversary of Uppsala University, and in August 1977 the Tenth International Congress of INQUA was held in Birmingham. The publications produced for these meetings contained material relevant to the content of this book. I am most grateful to Oxford University Press for allowing me to make slight alterations and additions to the text incorporating some of this new material at a very late stage in the production of the book.

M. J. TOOLEY

Witton-le-Wear
March 1978

CONTENTS

NOTE
There is no significance in the difference between 'metres O.D.' and 'm O.D.' as printed in the text.

LIST OF FIGURES

LIST OF TABLES

CHAPTER 1

Introduction

OBJECTIVES

There are extensive low-lying areas adjacent to the coast in the developed and developing world, that are close in altitude to the present-day sea-level. These areas are the focus for industrial and commercial activities, and are either densely populated or are recording immigration. The risks involved in developing such areas are high, when there is an unrealised potential global rise of sea-level of at least 50 metres, if all high latitude ice melted and the water returned to the ocean basins (Tooley 1971b, 1974).

The importance of sea-level studies lies not only in the prediction of the magnitude of sea-level movements, but also in estimating the extent of marine transgressions and regressions that might occur. Climatic change and sea-level movements are intimately related, and the direction and size of one, may be inferred from the other. Plant and animal migrations may be obstructed by a rise of sea-level, isolating continents and off-shore islands, and there can be little doubt that the poverty of the flora and fauna of the British Isles arose because land connections with the continent were severed about 8,500 years ago.

Sea-level studies have lacked an accepted methodology, and raw data collected in a variety of different ways have often been misinterpreted. Methods of data collection and analysis have varied in different parts of Europe, in North America and Australasia to such an extent that comparisons and meaningful correlations are difficult and lead to erroneous conclusions. To overcome these difficulties, a UNESCO-IGCP project on sea-level changes was initiated in 1974. The project working party agreed upon a standard form of return for sea-level data, drawing on the research experience of the last forty years, during which time different methods have been used in local and regional studies to establish the nature, size and time of sea-level changes. In general, the interdisciplinary approaches of von Post (1933, 1968), Iversen (1937), Godwin (1940a, b) and Jelgersma (1961) are preferable to those which concentrate on geomorphological or stratigraphic and radiometric analyses of a primary and derived nature, to the exclusion of all other approaches.

The main objectives of the UNESCO-IGCP project are to establish a graph of the trend of sea-level during the last 15,000 years and to predict sea-level trends in densely populated, low-lying coastal areas. The realisation of these objectives is through a series of detailed, local studies in the developed world, and the application of methods, tested in these studies, to other parts of the world.

The present piece of work fulfils the role of testing some of the methods that have been used to establish sea-level movements in an area of limited extent.

The work has two further objectives: the first is to present details of field and laboratory data from the coastal area of Lancashire, in a form sympathetic to the one proposed by UNESCO-IGCP working party, demonstrating the effectiveness of such an approach; the second is to synthesise these data into a regional scheme of marine transgression sequences, and to derive a relative sea-level curve. The argument is inductive, and stress is laid on the nature and quality of the field and laboratory observations and descriptions, as a basis for strong empirical models. These models require testing in hitherto unsampled areas, and may be used as a basis for tentative correlation elsewhere in Britain and Europe.

DESCRIPTION OF THE AREA

In 1956, Godwin wrote that, 'the most favourable conditions for determining former land- and sea-levels are in estuaries, where there has been a long inter-play between freshwater and marine brackish water deposits'.

It is fortunate that both North-west Europe (Fig. 1) and north-west England provide many such sites, where rivers and streams reach the coast, and are flanked by low-lying areas rarely in excess of +7.0 metres O.D.

In the original survey of the coasts of North Wales and north-west England, thirty-six sites were examined (Fig. 2) and described. But in order to fulfil one of the recommendations of the UNESCO-IGCP working party that sea-level studies should be in areas of limited geographical extent, the results from sampling sites 6–19, flanking the estuaries of the Mersey and Ribble (Fig. 2) are considered here.

Unity is given to the whole coastal area by the position of the coast along the southern and eastern margins of the east Irish Sea sedimentary basin (Bott, 1968). The basement rocks, for the most part, are Bunter and Keuper Sandstones, although marine sedimentation has been over Silurian and Carboniferous rocks in the Morecambe Bay area. The solid, however, is only occasionally visible at the coast, forming promontories: for example the Carboniferous limestone headlands of Arnside Park and Humphrey Head.

The intervening lowlands over the Triassic rocks have been filled with un-consolidated deposits of Quaternary age, contributed by successive glaciations that have affected the area. There is no unequivocal evidence from north-west England of sediments referable to a pre-Devensian stage, although the bio-genic deposits recorded between two till sequences in Furness, described by Bolton (1862) and Hodgson (1863) still need to be proved. Erosional features around Morecambe Bay, such as wave-cut notches and sea caves may be pre-Devensian (see review in Tooley 1977b). The till thicknesses given for north-west England may be referable only in part to the Devensian glaciation; maximum till thickness occurs in the Blackpool area, where 77 metres have been recorded. Terrestrial and marine sedimentation within basins since the end of the Devensian glaciation exceeds 25 metres locally: in Morecambe Bay, some 25–30 metres of Late Devensian and Flandrian sediments have been reported from the Leven Estuary. Sediments referable to the Flandrian stage alone exceed 16 metres in the zone affected by the Flandrian rise of sea-level in the Fylde of Lancashire.

The nature of sedimentation in areas adjacent to the present coast, where ground heights rarely exceed +7.0 metres O.D. has been affected fundamentally in north-west England by the rise of sea-level during the Flandrian stage. Within these low-lying areas, sedimentation under tidal, lagoonal or nearly marine conditions has been recorded. Hageman (1969) has made a distinction between the fossil sedimentary environments of coastal and near coastal areas of the Netherlands, and these distinctions have been applied wherever possible to the lithostratigraphic successions in north-west England. He distinguishes a lagoonal and tidal flat zone, in which full marine and estuarine facies alternate with biogenic facies often with a dominant saltmarsh element. A perimarine zone is defined as one in which marine and estuarine facies are absent, but continuous biogenic sedimentation is intimately related to movements of sea-level, which are registered in the perimarine zone by the elevation of the freshwater table, leading to the accumulation of gyttjas, freshwater clays and freshwater peats. The final division is the coastal barrier and dune area, the stability of which has been affected by movements of sea-level and changes of climate.

SUMMARY OF THE REGIONAL LITERATURE

The superficial geology of coastal Lancashire and southern Cumbria has interested local observers for the last three hundred years, particularly when new sections and cores revealed an unexpected stratigraphic succession or a stratum rich in faunal remains.

In the seventeenth century, the Revd. Richard James (1636) was aware of the differences in altitude between the spring tide levels and the low-lying mosslands of south-west Lancashire, that were susceptible to inundation. Furthermore, he described the intertidal peat beds off the coast of south-west Lancashire, and showed an awareness of relative sea-level changes.

In the nineteenth century, interest in the distribution and explanation of superficial deposits along the coast is shown by the field excursions, lectures and journals of the Liverpool and Manchester Geological Societies.

E. W. Binney and J. H. Talbot read a paper at the Annual General Meeting of the Manchester Geological Society, on 6 October 1843, *On the petroleum found in the Downholland Moss, near Ormskirk*, which was remarkable for its stratigraphic descriptions (see Chapter 2) and for its interpretation of the stratigraphic successions. They explained the marine transgressions on Downholland Moss in terms of the breaching of sand barriers, and the intercalating biogenic deposits in terms of shoaling, the creation of sand barriers, the blocking of tidal inlets and the restriction of free drainage from the mosslands. Such an explanation is similar to the one proposed by Jelgersma *et al.* (1970) to account for the pattern of marine transgressions in the Netherlands.

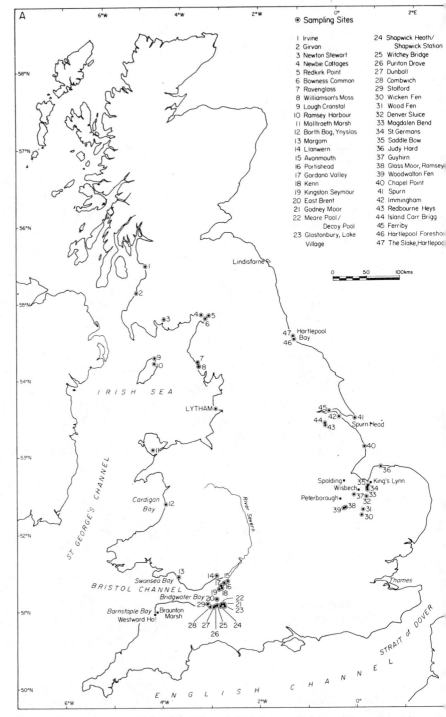

A

⊛ Sampling Sites

1 Irvine	24 Shapwick Heath/
2 Girvan	Shapwick Station
3 Newton Stewart	25 Witchey Bridge
4 Newbie Cottages	26 Puriton Drove
5 Redkirk Point	27 Dunball
6 Bowness Common	28 Combwich
7 Ravenglass	29 Stolford
8 Williamson's Moss	30 Wicken Fen
9 Lough Cranstal	31 Wood Fen
10 Ramsey Harbour	32 Denver Sluice
11 Malltraeth Marsh	33 Magdalen Bend
12 Borth Bog, Ynyslas	34 St Germans
13 Margam	35 Saddle Bow
14 Llanwern	36 Judy Hard
15 Avonmouth	37 Guyhirn
16 Portishead	38 Glass Moor, Ramsey
17 Gordano Valley	39 Woodwalton Fen
18 Kenn	40 Chapel Point
19 Kingston Seymour	41 Spurn
20 East Brent	42 Immingham
21 Godney Moor	43 Redbourne Heys
22 Meare Pool/	44 Island Carr Brigg
Decoy Pool	45 Ferriby
23 Glastonbury, Lake	46 Hartlepool Foreshore
Village	47 The Slake, Hartlepool

FIG. 1. Maps of the British Isles (A) and North-west Europe (B) to show the location of sampling sites described in the text.

20°W   10°W   0°   10°E   20E   30°E   70°N

66°N

62°N

N O R W A Y

S W E D E N

F I N L A N D

58°N

Goteborg
Viskan
Valley
Blekinge
Torekov   Hallarums Mosse
DENMARK   Siretorp
Great Belt
Prasto Fjord   54°N

Uitgeest
THE
NETHERLANDS
Ijmuiden   Velsen
Alphen
Brondwijk
(WEST)   (EAST)   P O L A N D

G E R M A N Y

50°N

BELGIUM

Camiers
St. Firmin
Tréompan   Le Havre
Argenton
Bréhec

C Z E C H O S L O V A K I A

F R A N C E

SWITZERLAND   A U S T R I A   H U N G A R Y

500kms

Fromentine
Brétignolles   Champagné
Aunis

46°N

Y U G O S L A V I A

P O R T U G A L

S P A I N

I T A L Y

42°N

A L B A N I A
G R E E C E

38°N

M E D I T E R R A N E A N   S E A

0°   10°E   20°E

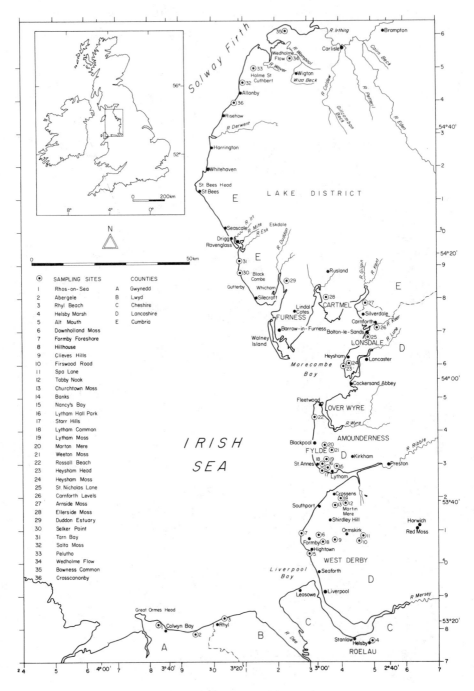

SAMPLING SITES

1	Rhos-on-Sea
2	Abergele
3	Rhyl Beach
4	Helsby Marsh
5	Alt Mouth
6	Downholland Moss
7	Formby Foreshore
8	Hillhouse
9	Clieves Hills
10	Firswood Road
11	Spa Lane
12	Tabby Nook
13	Churchtown Moss
14	Banks
15	Nancy's Bay
16	Lytham Hall Park
17	Starr Hills
18	Lytham Common
19	Lytham Moss
20	Marton Mere
21	Weeton Moss
22	Rossall Beach
23	Heysham Head
24	Heysham Moss
25	St. Nicholas Lane
26	Carnforth Levels
27	Arnside Moss
28	Ellerside Moss
29	Duddon Estuary
30	Selker Point
31	Tarn Bay
32	Salta Moss
33	Pelutho
34	Wedholme Flow
35	Bowness Common
36	Crosscanonby

COUNTIES

A	Gwynedd
B	Lwyd
C	Cheshire
D	Lancashire
E	Cumbria

The most substantial contributions towards a knowledge of the superficial deposits of coastal Lancashire came from T. Mellard Reade (1871, 1872, 1881, 1902 and 1908) and C. E. de Rance (1869, 1872, 1875, 1877 and 1878). Reade's (1871) cross-sections of south-west Lancashire (Fig. 14), show a lower peat, overlain by a marine clay (the Formby and Leasowe Beds), which is overlain in turn by an upper peat, blown sand and contemporary alluvium along the coast. Reade acknowledged that this was an over-simplification and noted that the upper peat was intercalated locally with marine clay.

De Rance argued that the Formby and Leasowe Beds of Reade contained a freshwater facies, which he defined as Lower *Cyclas* clay from the predominance of *Cyclas cornea* Linné (= *Sphaerium corneum* (Linné)). The marine facies was described as *Scrobicularia* clay after the type mollusc *Scrobicularia piperata* Gmelin (= *Scrobicularia plana* da Costa). *Sphaerium corneum* (Linné) has not been recorded in deposits of Flandrian age in southwest Lancashire since 1869, although living, or recently dead, specimens occur occasionally in drainage dykes (see Reade 1872), and, it is unfortunate that this subdivision of the *older estuarine and marine alluvium* has been retained by Evans and Arthurton (1973) as a basis for subdivision and correlation of the Flandrian stage in north-west England.

Further south, in Lancashire and Cheshire, Morton (1887, 1888 and 1891) has described similar stratigraphic successions adjacent to the estuary of the River Mersey with peats intercalating marine clays. Like Picton (1849) and Reade (1871), Morton explained these successions in terms of repeated land subsidence and elevation.

With this rich body of stratigraphical information, attention was directed during the first part of the twentieth century towards the plant micro- and macro-fossil content of deposits in south-west Lancashire. W. G. Travis (1908, 1922) described the macro-fossil plant remains from peaty horizons in the fossil sand-dunes around Aintree and Wallasey, whilst his brother, C. B. Travis (1926, 1929) carried out the first systematic examination of the plant, micro- and macro-fossil content of the coastal peat beds at the Alt Mouth and Leasowe. At the Alt Mouth, Travis recorded little change in pollen composition throughout the deposit. Contemporaneously, Erdtman (1928) visited this coast and carried out pollen analyses on samples taken from the coastal peat beds at Leasowe and the Alt Mouth in south-west Lancashire and from Ellerside and White Mosses in Southern Cumbria. Erdtman assigned the Leasowe peat to the Atlantic/Sub-Boreal Transition, the Alt Mouth peat to the early Sub-Boreal, and the culminating stage of the marine transgression beneath Ellerside Moss to the Atlantic, all of which have been confirmed by more recent pollen analyses and radio-carbon dates (see Chapter 2, and Oldfield and Statham 1963). In 1939, Blackburn published a short report (in

FIG. 2. A map of north-west England to show the location of the sampling sites, a selection of which is described in the text. The sampling sites have been grouped into counties, and Lancashire has been further sub-divided into the following four units: West Derby, Fylde in Amounderness, Over-Wyre in Amounderness and Lonsdale.

Cope 1939) on the pollen assemblages of the peat and diatom assemblages of the peat and clay on Downholland Moss, and was able to demonstrate that the deposits were of Atlantic age, and that the clays had accumulated under marine and brackish water conditions.

From 1934 until 1969, R. K. Gresswell published many contributions on Quaternary stratigraphy, geomorphology and coastal evolution during the Flandrian stage in Lancashire and southern Cumbria. His ideas on the evolution of the coast, undoubtedly influenced by W. B. Wright (1914), required that at the end of the last glaciation, sea-level was low, but, with the recession of the ice, sea-level rose and transgressed the Lancashire coast as far eastwards as the present 17 ft O.D. contour, where a sea-cliff was eroded in the boulder clay. A reduction of the eustatic rise of sea-level and continued isostatic recovery in Lancashire caused a relative fall of sea-level. Gresswell described the following distinctive assemblage of geomorphological features associated with these events: the wave-cut notch at + 17 feet O.D., called the *Hillhouse coastline*, the marine abrasion platform cut in boulder clay seaward of the notch, the Shirdley Hill Sand and Downholland Silt Formations. He correlated the *Hillhouse coastline* with the 25 foot raised beach in Scotland, and assigned it an age of 5,000 years BP. Gresswell mapped and recorded the altitude of the *Hillhouse coastline* in south-west Lancashire (1953, 1957), in the Fylde (1967) and in Lonsdale (1958). The *Hillhouse coastline* and the extent of the *Hillhouse Sea* are shown on Figure 14, according to Gresswell (1953, 1964). This model of coastal evolution in Lancashire and southern Cumbria has served as a basis for explanation and correlation for the last twenty years, and coastal sequences are referred to Gresswell's scheme (Andrews *et al.* 1973; Corlett 1972; Evans and Arthurton 1973; Hall 1956; Hall and Folland 1970; Stephens and Synge 1966; Walker 1966; Whittow 1965, 1971). However, as early as 1918, Pearsall had mapped the post-glacial limit of the freshwater lake in Rusland Valley, and Dickinson (1973) has shown that this limit coincides very closely with the raised, marine beach that Gresswell mapped in 1958. Furthermore, in 1928, it would have been possible, by collating the published stratigraphic and palaeobotanic data from Leasowe and Ellerside Moss (Reade 1871; Erdtman 1928), to give an estimate of 2.6 metres of isostatic recovery in southern Cumbria, since the Atlantic, which compares with a negative value of 0.53 metres given by Gresswell (1958) and subsequently transformed by him to a positive value of 0.38 metres. In south-west Lancashire, Binney and Talbot (1843), Reade (1871), Cope (1939) and members of the Soil Survey of England and Wales (Hall, 1954-5) had proved the complexity of the stratigraphic sequences on the lowland mosses, that had been explained inadequately by Gresswell. Godwin (1959) had demonstrated the late-glacial age of the Shirdley Hill Sands, in contradiction to the mid-post-glacial age and coastal origin assigned to them by Gresswell (1953), Wray and Cope (1948) and Evans and Arthurton (1973). Further north, in coastal areas of Over-Wyre, Lonsdale and southern Cumbria, detailed stratigraphic and pollen analytical studies made the reconciliation of both the concept and reality of the *Hillhouse coastline* with the results of these studies progressively more difficult.

More recent, local palaeobotanical and stratigraphic studies include work by Dickinson (1973, 1975), Oldfield (n.d., 1958, 1960, 1963 and 1965), Oldfield and Statham (1963, 1965), Moseley and Walker (1952), Huddart *et al.* (1977), Tooley (1969, 1970, 1971a, 1973, 1974, 1977a) and Smith (1958b, 1959). Further reference is made to these and other papers in the succeeding chapters.

TECHNIQUES

Many techniques have been used during the last hundred years to establish the pattern and age of sea-level changes. There has been a tendency for workers to concentrate on a few, selected techniques, such as stratigraphic and radio-metric analyses and plane surveying. Interpretations, based on the field data deriving from the application of such techniques, are open to discussion, because the nature of the palaeoenvironments has not been established or the radiometric analyses have not been substantiated by independent pollen analytical methods or magnetic intensity studies. The controversy between C. E. de Rance and T. Mellard Reade in the latter part of the nineteenth century arose from a difference of interpretation of the lithostratigraphic units in south-west Lancashire, based on the presence or absence of particular indicator fossils studied by each author. The beginning, maintenance and end of marine conditions in coastal areas during the Flandrian Stage can be established and clarified by the application of a range of physical, chemical and biological techniques. In 1933, H. Godwin and M. E. Godwin (1933a) and Macfadyen (1933) had used pollen, stratigraphic and foraminiferal analyses at St. Germans' in the Fenlands to elucidate the nature of the marine transgressions recorded there. Iversen, in 1937, indicated the potentialities of such an approach by applying stratigraphic, pollen and diatom analyses to lake deposits in Sobörg Sö in north Zeeland, Denmark and establishing four distinct transgressive phases during the Atlantic and Sub-Boreal periods.

Thus, when this work began in 1966, there was already a tradition of applying a range of techniques to elucidate pattern and age of sea-level changes, and, with refinement, the following six techniques were used, levelling, stratigraphic analysis, pollen analysis, diatom analysis, chemical analysis and radiometric analysis.

A summary of three of these techniques is given below, together with a consideration of the nature of the sources of error that are likely to affect both the field and laboratory data, and thereby the strength of the interpretations and subsequent conclusions. Further details are given in Tooley (1969, in the press).

Levelling

A Kern Automatic Level, GKI-AC, was used to obtain the altitudes of the sampling sites. Two or more bench marks were used to obtain the reduced level of the ground at each sampling site. The positions and altitudes of bench

marks were obtained from the Ordnance Survey Bench Mark Lists, all of
which derived from the Third Geodetic Levelling of the United Kingdom.
The closing errors on the heights of the sampling sites ranged from 0.002 to
0.170 metres. All measured altitudes are related to Ordnance Datum
(Newlyn), which is the zero datum of the United Kingdom, and is equivalent
to the average value of mean sea-level at Newlyn, Cornwall for the six-year
period 1915–1921. Since 1921, Ordnance Datum (Newlyn) has moved relative
to sea-level, and is 0.1 metres below mean sea-level (Admiralty Tide Tables
1970).

Stratigraphic Analysis

The stratigraphic descriptions given in Chapter 2, and the symbols (Fig. 3)
displayed in the figures derive from records taken in the field, in part with a
Beus and Mattson Hiller-type peat sampler, a Russian-type peat sampler and
a Duits' gouge sampler, in part from monolith samples taken from free-face
excavations, and in part from the borehole logs of consulting engineers. An
attempt has been made to standardise the stratigraphic descriptions and
symbols, according to the scheme proposed by Troels-Smith (1955). Accordingly, unconsolidated deposits are characterised by:
  (a) the elements of which they are composed,
  (b) the degree of humification of the organic element(s), and
  (c) the physical nature of the deposits.
Each group of characteristics has the advantage of assessment in the field, on
a five point scale: 4 indicates maximum presence of an attribute, whereas 0
indicates absence. Further elaboration is possible after the samples are
returned to the laboratory, but essentially, the descriptions given here derive
from field records, and the results of laboratory analyses are given separately.
  The stratigraphic descriptions given in the text are presented in a
standardised format: each stratum identified is assigned a number; the height
of the stratum is given in metres in relation to Ordnance Datum (Newlyn); the
depth from the surface is shown in centimetres; and a description of the
sediment type is given both in an abbreviated formula after the notation of
Troels-Smith (1955) and in a conventional description. The abbreviated
formula contains, on the first-line, a list of all the elements present, their
proportions and degree of humification, if biogenic. The second line of the
description gives a list of physical characteristics of the stratum: that is, the
degrees of darkness (nig. = *nigror*), stratification (strf. = *stratificatio*),
elasticity (elas. = *elasticitas*) and dryness (sicc. = *siccitas*). Finally, an estimate
is made of the boundary type of each stratum: the upper boundary (lim. sup.
= *limes superior*) is specified in most cases on the scale lim. 4 acute, to lim. 1
diffuse. At Ansdell −2, stratum 3 is shown by the abbreviated formula as:

FIG. 3. A key to the main groups of symbols used in the stratigraphic diagram. The superscript
numbers indicate degrees of humification of the element recognised. Further details are given
by Troels-Smith (1955).

KEY TO THE STRATIGRAPHIC SYMBOLS
after TROELS—SMITH (1955)

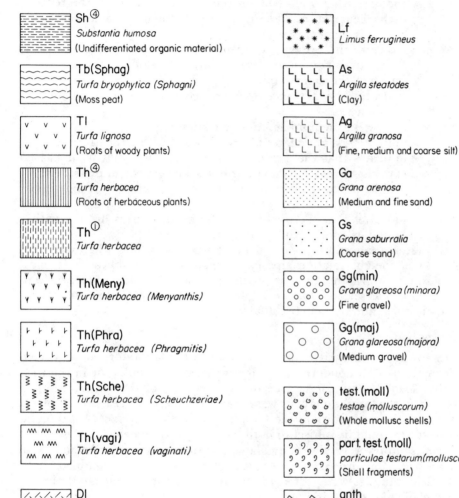

Sh④
Substantia humosa
(Undifferentiated organic material)

Tb(Sphag)
Turfa bryophytica (Sphagni)
(Moss peat)

Tl
Turfa lignosa
(Roots of woody plants)

Th④
Turfa herbacea
(Roots of herbaceous plants)

Th①
Turfa herbacea

Th(Meny)
Turfa herbacea (Menyanthis)

Th(Phra)
Turfa herbacea (Phragmitis)

Th(Sche)
Turfa herbacea (Scheuchzeriae)

Th(vagi)
Turfa herbacea (vaginati)

Dl
Detritus lignosus
(Wood and bark fragments)

Dh
Detritus herbosus
(Stems and leaves of herbaceous plants)

Ld④
Limus humosus
(Fine detritus mud)

Lf
Limus ferrugineus

As
Argilla steatodes
(Clay)

Ag
Argilla granosa
(Fine, medium and coarse silt)

Ga
Grana arenosa
(Medium and fine sand)

Gs
Grana saburralia
(Coarse sand)

Gg(min)
Grana glareosa (minora)
(Fine gravel)

Gg(maj)
Grana glareosa (majora)
(Medium gravel)

test.(moll)
testae (molluscorum)
(Whole mollusc shells)

part. test. (moll)
particulae testarum (molluscorum)
(Shell fragments)

anth
anthrax
(Charcoal)

Str. Conf
stratum confusum
(Disturbed stratum)

P
exemplum stratorum
(Samples taken for pollen analysis)

D1.1. Th^23. Th^1 (*Phra*) + . Gs + .

nig. 2, strf. 1, elas. 0, sicc. 2, lim. sup. 0

This formula is derived in the following way:

D1.1	indicates that 25% (=1) of the stratum comprises woody, detrital peat. The woody element has not been identified in the field.
Th^23	indicates that 75% (=3) of the stratum comprises the roots of herbaceous plants. The superscript 2 indicates a humification of 2 for this element.
Th^1 (*Phra*)+	indicates the presence (=+) of rhizomes of the common reed, *Phragmites communis* Trin., and the slight humification is shown by the superscript.
Gs +	indicates the presence (=+) of coarse sand throughout the stratum.
nig. 2	indicates a medium degree of darkness.
strf. 1	indicates a slight degree of stratification within the stratum.
elas. 0	indicates that the stratum has no elasticity, that is, it has no capacity to regain its shape after the application of pressure.
sicc. 2	indicates that the stratum is saturated with water, at the time of observation.
lim. sup. 0	indicates that the upper limit of the stratum is diffuse, with a boundary zone greater than 1 cm.

Stratigraphical analyses were carried out at sampling sites where the geo-graphical and height relationships had already been established. In most cases, a full stratigraphic analysis of the site was completed, before a rep-resentative site was selected for the acquisition of a complete core for pollen, diatom, chemical and radiometric analyses.

Pollen Analysis

Samples for pollen analysis were taken either in the laboratory directly from the cleaned faces of monoliths, or half-core Russian samples, or in the field from the cleaned sediment surface in the chamber of the Hiller. Samples deriving from the former methods were preferred, because greater control was possible over their acquisition, the sample thickness rarely exceeded 0.5 cm, the same strata could be subject to both pollen and radiometric analyses, and additional samples were always available for further analysis if elaboration was necessary. Monolith samples were sealed in polythene and complete, half-core Russian samples were stored in 50 cm plastic tubes split longitudinally and the whole also sealed in polythene. Samples were taken directly from the chamber of the Hiller with 1 cm diameter glass vials, distilled water was added to minimise dessication, and the corks sealed with paraffin wax. Storage has been in a cold-room, maintained at 0 °C.

A standard laboratory technique was applied to the samples, the details of which are outlined elsewhere (Tooley 1969), but are based essentially on the practices described in Faegri and Iversen (1964) and Gray (1965). The technique had to be adjusted to accommodate the variety of sediments that were subjected to the method. The isolation of micro-fossils from biogenic sediments sampled between 1966 and 1969 has been carried out by myself, while samples acquired between 1969 and 1972 have been analysed by Mr J. Telford. The identification and enumeration of pollen taxa has been undertaken by myself.

The percentage frequencies of the pollen taxa identified are presented in a more or less standardised form in the pollen diagrams: there are slight differences between pollen diagrams presented in my thesis (Tooley 1969) and those that have been drawn up subsequently. However, all pollen frequencies that derive from the former have been recalculated to conform with the latter pattern. The pollen taxa identified have been grouped into the life-form classes of the plants producing the pollen: the A or AP Group comprises all tree taxa identified, Group B — Coryloid, Group C — Shrub taxa, Group D — Herb taxa, Group E — Aquatic taxa, Group F — Pteridophyte and Bryophyte taxa, and miscellaneous groups which comprise unidentified taxa, *Pediastrum*, Dinoflagellate cysts shown as Hystrichosphaeroidae and fungal spores such as *Tilletia sphagni*. In some diagrams, Coryloid numbers are included with the tree taxa and subsequent groups relettered accordingly. The minimum number of Group A or AP counted was 150: a minimum of 200 was preferred, but some of the earlier diagrams used the lower sum, and the paucity of tree taxa at other sites prevented the realisation of this minimum count, and in this case a sum of 500 TP was counted. The percentage frequency is calculated for each taxon at successive levels according to the calculating formula given for each group, which is quite simply a formula for calculating percentages. The denominator for each group comprises the sum of the A or AP Group and all the taxa identified and assigned to that group at successive levels: in this way percentage frequencies in excess of 100 are avoided.

The pollen diagrams have been subdivided into one or more pollen assemblage zones, specified by the initials of the sampling site and an appropriate subscript letter. The criterion for subdivision has been the behaviour of the AP Group taxa, and boundaries between pollen assemblage zones have been drawn where the percentage frequency of one or more taxon changes significantly at successive levels. Local pollen assemblage zones, which refer to a pollen diagram at the sampling site, are shown on the pollen diagrams.

There were two main purposes for constructing pollen diagrams:

(a) to elucidate the changing environmental conditions within the tidal flat, lagoonal and perimarine zones during the period of biogenic accumulation, and

(b) to serve as a check on the radiocarbon dates that were obtained from organic material adjacent to the contacts of marine facies.

An attempt to establish regional pollen assemblage zones and chronozones at Marton Mere — a low-lying kettle hole, adjacent to the perimarine zone, and geographically central to the sites considered — was abandoned when the site was covered by dumping. The pollen diagram from 750 cm of the mere sediments shows a complete Late Devensian and Flandrian vegetational sequence (Tooley 1977a).

The difficulty in corroborating radiocarbon dates, by referring local pollen assemblage zones to the regional pollen assemblage zones and chronozones at a type site, was overcome when Hibbert *et al.* (1971) adopted Red Moss (SD 631103) as the Flandrian type site in north-west England (Fig. 2).

Three chronozones have been defined at Red Moss: Flandrian I is a pre-temperate stage dominated by boreal trees, such as *Betula* and *Pinus* but with increasing proportions of thermophilous tree taxa, which serve as a basis for the subdivision of the Chronozones a, b, c and d. Flandrian II is an early-temperate zone in which mixed oak forest dominates the vegetation, par-ticularly, *Quercus*, *Ulmus*, *Fraxinus* and *Corylus*. Flandrian III is equivalent to the late-temperate zone recognised by Turner and West (1968) in the suc-cession of interglacial stages. It is characterised by the immigration of temperate tree taxa such as *Carpinus* and *Fagus*, except that in the Flandrian it is overwhelmingly the zone in which vegetational composition and patterns are affected by human activity (West 1970).

The absolute time limits for the regional pollen assemblage zones recognised at Red Moss are given in tabular form below:

Chronozone	Date, years BP	Regional Pollen Assemblage Zones (RPAZ)		
FIII	5010±80 to present	*Quercus*	— *Alnus*	
FII	7107±120 to 5010±80	*Quercus*	— *Ulmus*	— *Alnus*
FId	8196±150 to 7107±120	*Pinus*	— *Corylus*	— *Ulmus*
FIc	8880±170 to 8196±150	*Corylus*	— *Pinus*	
FIb	9798±200 to 8880±170	*Betula*	— *Pinus*	— *Corylus*
FIa	? to 9798±200	*Betula*	— *Pinus*	— *Juniperus*

In the pollen diagrams presented here, the Local Pollen Assemblage Zones identified are correlated with the Regional Pollen Assemblage Zones at Red Moss. Where there is sufficient evidence, that is, a radiocarbon date, the validity of which is corroborated by concurrence of the local and regional pollen assemblage zones, correlation between the local site and the type site has been attempted.

Fig. 4 shows the temporal pattern of the Flandrian Chronozones, and the correlation between the local pollen assemblage zones and the regional pollen assemblage zones at Red Moss, Horwich, Lancashire established by Hibbert *et al.* (op. cit.)

SOURCES OF ERROR

The techniques employed to elucidate problems of land- and sea-level relationships have provided a range of data which is subject to many sources of

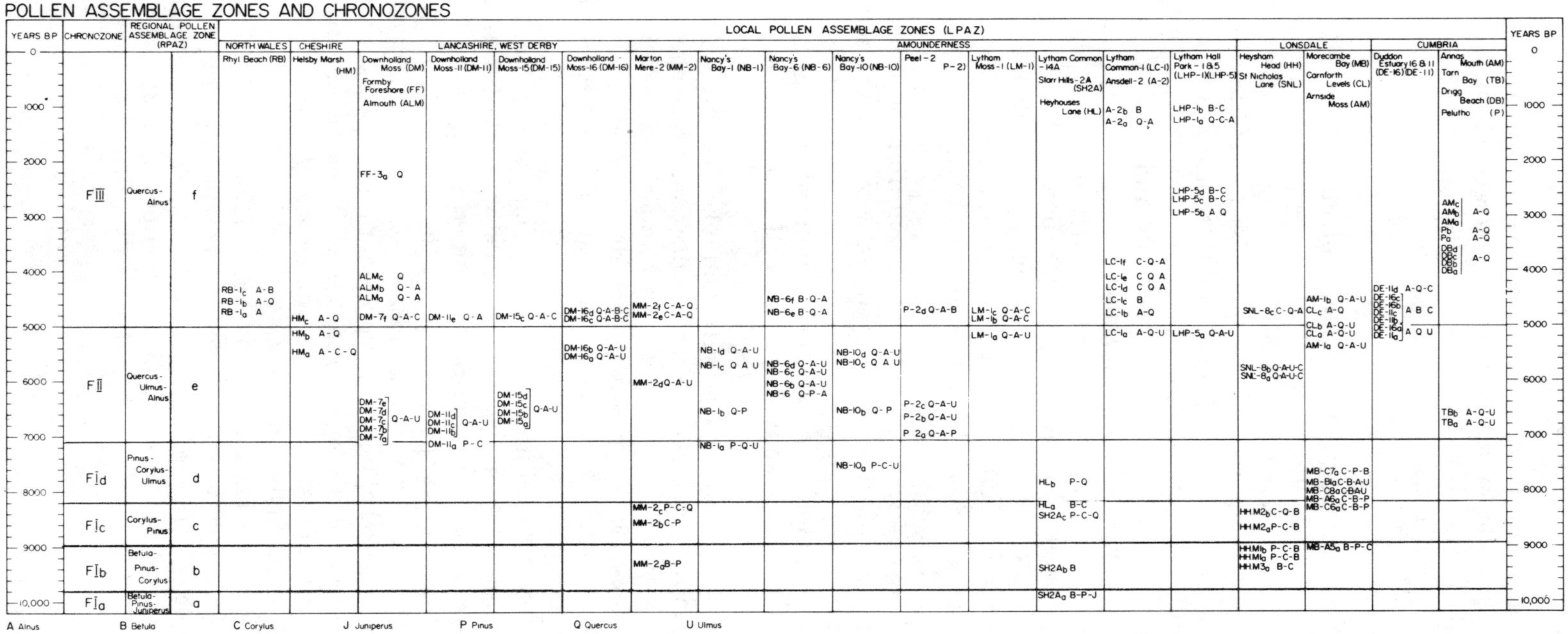

POLLEN ASSEMBLAGE ZONES AND CHRONOZONES

FIG. 4. Local and Regional Pollen Assemblage Zones and Chronozones from north-west England. The Chronozones and Regional Pollen Assemblage Zones are taken from Hibbert *et al.* (1971). Local Pollen Assemblage Zones are referred to the type site at Red Moss, and are located within chronozones when absolute dates permit.

error. The interpretation of these data is only possible if cognisance is given to the nature and magnitude of these sources of error, which include the differential consolidation of sediments, tectonic warping, tidal range, and the validity of the radiocarbon dates.

Consolidation of Sediments

All sediments undergo consolidation, the rate of which is a function of time, drainage and load, either under their own weight or from overlying sediments.

Models have been drawn up, which make it possible to allow for the amount of settlement that will occur when a specific load is applied to a homogeneous material whose properties are known. Terzaghi and Peck (1967) have considered the behaviour of a range of materials under stress, and have touched on the consolidation and compaction of marine clays and peats. Jelgersma (1961) has reviewed some of the literature on the consolidation of sediments, and stressed that consolidation of deposits with a high sand fraction is very low, whilst consolidation of peat may be as high as 90 per cent by volume. As there is a great variety of unconsolidated sediments in Flandrian coastal deposits, consolidation can vary from 0 to 90 per cent (Jelgersma, op. cit., p. 16).

The behaviour of peat under stress is quite different from any other unconsolidated sediment, and MacFarlane (1965) has described the settlement characteristics of peat as a function of time:

(a) Primary consolidation occurs extremely rapidly when a load is applied, and this consolidation can account for up to 50 per cent of the total settlement.

(b) Secondary consolidation occurs slowly, and takes the form of a type of viscous or plastic movement especially if the water content changes or decay accelerates. The length of time taken for secondary consolidation is not known, but Barden (1968) notes that, 'certain peats appear to give a secondary settlement linear with log-time, that appears to extend indefinitely, although it must cease eventually'.

Different types of peat have different types of properties, which affect the period of consolidation. Unhumified fibrous peats have an open structure, which will hold more interstitial water than humified peats: the void ratio is also higher and this affects the permeability of the peat and therefore the response to changes in ground water conditions.

As the biological processes of decay take place within the peat there will be a loss of volume. The rate of decay is a function of water content, the availability of oxygen, the temperature and the rate of peat growth, all of which change continuously in the body of the peat.

MacFarlane (op. cit.) has also noted that there appears to be a significant amount of rebound during or following secondary consolidation of peat, if the applied load is reduced or removed. Indeed, one of the physical properties for which allowance is made in Troels-Smith's description of unconsolidated sediments, is the degree to which a sediment can regain its original form after stress has been applied to it, i.e. the degree of elasticity. Thus, fresh *Sphagnum* peat will more or less regain its shape after pressure has been applied and released, whereas woody, detrital peat has no capacity to regain its shape. Sites in north-west England that have probably suffered from rebound during or following secondary consolidation of peat include the Alt Mouth, Lytham Common and Lytham Hall Park. Along the coasts of Lancashire and Cumbria, the coastal peat has been obscured by an over-burden of dune sand, which locally exceeds a thickness of 30 metres: at Lytham Common-1, there is an overburden of sand 1.4 m thick, which began to accumulate there at 805±70 BP. At the Alt Mouth, the elliptical cross-sections of *Betula* sp. testify both to former loading by dunes that have now moved landward and to autoconsolidation.

Skempton (1970) and Greensmith and Tucker (1971b) have demon-strated that marine clays, silts and sands reveal varying degrees of over-consolidation, that is, they have been subjected to a pressure greater than the present load. Over-consolidation can result from either denudation during a cycle of erosion or dessication, which is equivalent to consolidation under load (Terzaghi and Peck 1967). Both these processes have operated in coastal areas affected by periods of relatively high and low sea-level during the Flandrian stage, and will affect the mechanical properties of marine deposits as Greensmith and Tucker (1971a, 1971b, 1973) have shown for the tidal flat and lagoonal zones of the Essex coast.

Sediments encountered in coastal areas are rarely homogeneous in vertical section and geographical extent. They comprise bryophyte and mono-cotyledonous peats showing a range of degrees of humification, woody detrital peats and gyttjas with a varying inorganic fraction, all of which can be separated by estuarine and marine clays, silts and sands. Each stratum thus distinguished will behave differently under stress, and any correction factor must take into account this diversity of sediment types in the stratigraphic column.

Although the computation and application of correction factors would be an impossible task for sediments from coastal areas exhibiting different degrees of consolidation, the importance of consolidation to a consideration of absolute movements of sea-level is paramount.

Jelgersma (op. cit.) has overcome the problem by dating and measuring the heights of peat that accumulated on Pleistocene sands. She argues that peat growth was causally related to the rise in sea-level, and that subsequently the peat has not been subject to significant consolidation. Elsewhere in the tidal flat and lagoonal zone of the Netherlands, she has dated peat of Calais and Dunkirk age which have been subject to consolidation and settlement for which no allowance has been made. At Brandwijk, between the distributaries of the Rivers Rhine and Meuse, south-east of Rotterdam, Jelgersma (op. cit., p. 32) has demonstrated that the thickness of the lower fen peat and fluvial clay has been reduced from 4.10 to 1.40 m as a result of loading and changes in ground-water conditions. Such an estimate is based on a series of levelled heights and radiocarbon dates, and refers only to site conditions at Brandwijk: extrapolation to other sites with a similar stratigraphic and temporal pattern is not valid.

With reference to the application and validity of these data to coastal studies, Jelgersma makes a fundamental distinction between data that can be used in a sea-level curve, and data that can only be used in an elucidation of coastal sequences, which for their spatial and temporal patterning do not depend exclusively on a consideration of consolidation.

A similar distinction has been made by Bloom (1964) who has estimated a consolidation factor for sedge peat, buried beneath fossil marine sediments in the Hammock River Valley, Connecticut. The factor has been derived by comparing the age and depth of peat samples from the peat/clay contacts with the curve for the rising high-tide level, established by dating the sedge peat at the contact with the underlying sands and gravels.

Both Jelgersma and Bloom assume a causal relationship between the inception of peat growth on a sloping Pleistocene sub-surface, rising ground water and approaching marine conditions as sea-level rises, even though the macro- and micro-fossil assemblages of the peat may give no indication at all of the proximity of marine conditions at the sampling site. The relationship of peat growth to sea-level then becomes a fundamental problem.

In the present study, no attempt has been made to derive factors empirically for the consolidation of sediments in the coastal area, and it is probable that height data of transgressive sequences have been affected by different amounts of consolidation for the reasons cited earlier. However, the cross-sections from West Derby and Amounderness (Figs. 33 and 34) showing the transgression sequences there give no evidence of appreciable consolidation and settlement or dislocation of the marine transgression surfaces.

The heights of dated samples in relation to Ordnance Datum have been plotted directly onto the Time-Depth graph (Fig. 36) and on to the sea-level curve graphs (Figs. 37 and 38). In Chapter 4, a qualitative estimation is made of the extent to which the index points used to construct the sea-level curve (Fig. 38) have been affected by consolidation.

The Relationship between Saltmarsh, Reedswamp and Fen Communities and Sea-Level

The interpretation of the lithologic boundaries in stratigraphic successions within the coastal zone may contribute a significant source of error, particularly when the effective height of a marine transgression is under consideration. The error arises from the fact that the boundary zone may occupy a whole stratum in the stratigraphy as the approaching marine conditions gain ascendancy at the site, but before inorganic sedimentation begins. The interpretation of the facies changes within the strata of clay, silt or sand is also elusive, and it is difficult to estimate the relationship to tidal levels during the period of accumulation and the rate of sedimentation. An attempt has been made in the case of Downholland Moss — 15 (see Tooley 1977a and in the press). As the stages of sedimentation in minerogenic strata are rarely susceptible to accurate dating, except where bivalve molluscs occur *in situ* in their living positions, the problem turns on the interpretation of the boundaries or boundary zones flanking these strata, for they usually contain sufficient organic material for a radiocarbon assay. Furthermore, an assessment of the micro- and macro-fossil content of these zones permits a tentative palaeo-ecological interpretation, and isolates the stratum which is most appropriate for a radiocarbon assay. If this procedure is omitted, then samples may be submitted which date a completely irrelevant geological or ecological event.

Jelgersma (1961) has dated and recorded the height of fen wood peat that had accumulated on Pleistocene sands, and has used these dates in the graph of relative changes in sea-level for the Netherlands. She has argued cogently that the peat formed slightly above the ground-water table, which is intimately related to sea-level. In the case of the Rhine–Meuse estuary she argues that fen wood peat would accumulate from 50 cm above to 50 cm below high-tide level. When undisturbed samples of fen wood peat are taken for radiocarbon assays from gently sloping sites on Pleistocene sands, the objections to sediment consolidation are overcome and data for a sea-level curve are obtained. However, in accepting the former, the latter is open to question. At Alphen Aan de Rijn, for example, the advent of marine conditions is a metre above, and seven hundred years after, the dated level used as an index point in the sea-level curve for the Netherlands (Jelgersma, op. cit., p. 83).

There is little published work available on the contemporary relationships between sea-level, soil conditions and the succession of coastal plant communities. However, such work is fundamental before coastal palaeo-envrionments can be interpreted adequately.

On the north side of the Ribble estuary at Lytham (Fig. 2), Cheesbrough *et al.* (1969) have described a *Puccinellia* marsh growing from +3.90 metres to +4.90 metres O.D., and a *Spartina* meadow with an upper limit of +4.45 metres and a lower limit of +2.89 metres O.D. Mean High Water (M.H.W.) at St. Annes is +3.30 metres O.D. and Mean High Water of Spring Tides (M.H.W.S.) reaches +4.42 metres O.D. On Lytham Marsh, further up the Ribble estuary at Lytham, the measured altitude of the drift line, interpreted as M.H.W.S.T. is at +4.82 metres O.D. Further north, on the eastern side of Morecambe Bay, on the Silverdale, Arnside, Warton and Bolton-le-Sands marshes, the grazed marshes of *Festuca* and *Puccinellia* are maintained between the limits +5.1 to +6.3 metres O.D. (Kerr 1967), and well above M.H.W.S. at Heysham which is +4.57 metres O.D. These altitudinal limits have been confirmed by Gray and Bunce (1972) who have identified four marsh types in Morecambe Bay. Mature marshes have an altitudinal range of +5.3 to +6.6m O.D.; high-level saltings have a range of +4.9 to +5.8 m O.D.; low-level saltings have a range of +4.5 to +4.7 m O.D. and pioneer zones have a range of +3.5 to +4.6 m O.D. Gray and Bunce (op. cit.) suggest that the formation of a salt marsh sward of *Puccinellia maritima* (Huds.) Parl. occurs at a critical altitudinal threshold in the low-level saltings range. Within this marsh type, only a narrow vertical difference in altitudes results in considerable differences in submergence: the pioneer-zone has a mean submergence of about 350 tides, wheras the low-level saltings have a mean submergence of about 200 tides and the high-level saltings about 50. This critical threshold has an altitude rather close to M.H.W.S. at Heysham, and salt marsh communities are maintained to two metres above this altitude in the mature marshes, on which six tidal submergences are recorded each year. If this relationship persisted throughout the Flandrian stage in Morecambe Bay, then the assumption (p. 22) that minerogenic sedimentation gives way to biogenic sedimentation at about M.H.W.S. may contain an error approaching this magnitude.

Most of the Lancashire marshes are grazed, and the landward margins have been reclaimed. There are no well-developed reedswamp communities and the succession to fen is missing. Landward of the Alt Mouth, at the foot of the dunes, and utilising the seepage of freshwater at the peat/dune sand interface, is a small reedswamp community dominated by *Phragmites communis* Trin. with some *Scirpus maritimus* Linné. The rhizomes and rootlets of the *Phragmites* form a fibrous mat of sandy peat, which is a distinctive facies of the high intertidal zone. This community occurs at a height of +4.0 metres O.D. which is 15 cm below M.H.W.S. at Formby.

In Scotland at the heads of some of the sea lochs on Mull, Gillham (1957) has described the relationship of the contemporary coastal plant communities to M.H.W.S. On Loch Scridain, a mixed grass, sedge and rush community was recorded from 23 to 59 cm above M.H.W.S., and was replaced from 59 to 83 cm above M.H.W.S. by a mixed moorland community, dominated by *Molinia caerulea* (Linné) Moench, *Eriophorum angustifolium* Honck and *Juncus articulatus* Linné, with occasional shrubs of *Myrica*. On both Loch Scridain and Loch Guin, within 15 metres of the shore, deep peats had accumulated and supported vegetation of *Myrica* and *Molinia* or *Calluna vulgaris* (Linné) Hull and *Erica tetralix* Linné. These latter communities occurred no more than 96 cm above M.H.W.S.

In Barnstable Bay, New England, Redfield (1967) has described the plant communities and the type of peat formed in the intertidal zone. High marsh peat accumulates at mean high water, and is composed of the rhizomes, stems and leaves of *Spartina patens*, *S. alterniflora* and *Distichlis spicata*, which can withstand limited submergence. The middle and upper intertidal zone is covered by *S. alterniflora*, and peat growth in this community has been estimated at 61 cm/century (Chapman 1960).

The succession from high saltmarsh, through reedswamp communities to shrub communities has been carefully recorded from Island Beach, New Jersey, by Martin (1959), in relation to mean sea-level, ground-water level and soil conditions. Salt marsh communities, dominated by *Spartina patens*, are maintained up to 34 cm above mean sea-level, and usually produce a peat 30 cm thick. Reedswamp communities occur 122 cm above sea-level, and ground water rarely falls below 30 cm from the surface. They are dominated by *Typha* and *Phragmites*, and are replaced by a scrub of *Juniperus, Myrica* and *Rhus* when the ground rises to 152 cm above mean sea-level.

It is not known at what level fen communities are maintained in relation to mean sea-level or M.H.W.S., or whether there is any relationship between the level of ground water in the fen and sea-level. Godwin *et al.* (1935) mention that fen woodland flourishes when the water table is at, or above, the ground surface, and one of Jelgersma's basic assumptions (op. cit., p. 21) is that fen wood peat in the Rhine-Meuse delta was formed 50 cm±high tide level. Kaye and Barghoorn (1964) describe two situations in which barrier beaches protect marshes, occasionally flooded by high spring tides, and where both *Salix* and *Pinus rigida* Miller flourish. Many other tree taxa have been reported flourishing at or close to M.H.W.S.: Richardson (in Jones, 1959, p. 179) reports that *Quercus robur* Linné survives inundation by sea water, and can withstand exposure to sea winds, although under these conditions it forms low scrub.

The most comprehensive study of the relationship of plants to sea-level is by Johnson and York (1915). Their detailed examinations of the marshes surrounding Cold Spring Harbour have given precise altitudinal data for the plant communities that they identified. The data they describe on supra-littoral vegetation are particularly relevant here. The supra-littoral zone occurs from 2.43 metres to over 3.04 metres above Mean Low Water. The zone is therefore affected by Spring tides, which rise, on occasions, to more than 3.35 metres above M.L.W. The vegetation is characterised by stands of *Alnus incana* (Linné) Moench. 3.65 metres high, shrubby communities of *Baccharis* sp., *Hibiscus moscheutos, Iva* sp. and *Solanum dulcamara* Linné, localised clumps of fresh-water marsh plants such as *Iris versicolor* Linné and occasional individuals of *Acer rubrum* Linné, *Fraxinus americana* Linné, *Juniperus virginiana* Linné, and *Tilia americana* Linné. In this zone, the plant communities are subject to submergence during the spring tide cycle from May to October on 10 to 65 occasions, and the period of immersion lasts from 2 to 60 minutes. As these tree taxa, some of which are components of fen communities, are growing at or below the spring tide mark, it is unlikely that the factor of 50 cm ± high tide level that Jelgersma uses for the level at which fen wood peat accumulates, has great validity: at Cold Spring Harbour, fen wood peat is accumulating some 80 cm below the spring tide level and reedswamp peat dominated by *Typha angustifolia* Linné with some *Iris versicolor* Linné occurs from 70 cm to 200 cm below the spring tide level.

The altitudinal range of peat-forming plant communities in the intertidal zone makes the interpretation of boundaries and boundary zones between biogenic and minerogenic strata difficult. Local factors, such as barrier beaches, shore progradation, lagoons and changing sedimentary régimes in estuaries may have a profound effect on the extent and composition of littoral vegetation. The assignment of a single correction factor to all dated boundary heights will therefore add to the errors rather than reduce them.

Tidal Range

The greatest, mean tidal ranges in the British Isles are experienced in the Bristol Channel and along the Lancashire coast, where the range exceeds six metres.

At Liverpool, mean high water (M.H.W.) is at +3.46 metres O.D. and mean low water (M.L.W.) at -3.01 metres. On the open coast, at St. Annes, M.H.W. is at +3.30 metres and M.L.W. at -2.35 metres, and in Morecambe Bay, Heysham records M.H.W. at +3.55 metres and M.L.W. at −2.78 metres (Admiralty Tide Tables 1967). At St. Annes, the range of the spring tides is 7.96 metres and at Heysham 8.38 metres. Such a range has produced a sandy foreshore during mean low water more than 5 km wide off St. Annes and Southport, and much of Morecambe Bay is evacuated of water during low spring tides.

As it is assumed that mean high water of spring tides is the effective level about which salt marsh communities and reedswamp communities occur, the actual heights of contemporary high water marks along the Lancashire coast are given below, in metres in relationship to Ordnance Datum:

	M.H.W.S.	M.H.W.N.	M.H.W.
Formby	+4.15 m	+2.38 m	+3.26 m
Liverpool	+4.40 m	+2.52 m	+3.46 m
St. Annes	+4.42 m	+2.19 m	+3.30 m
Heysham	+4.57 m	+2.53 m	+3.55 m

It is assumed also, that the tidal range has remained more or less constant along the Lancashire coast during the Flandrian stage, the same assumption made by Jelgersma (op. cit., p. 22) for the Netherlands' coast. It is admitted, however, that the change in the coastline around the Irish Sea during stages in the restoration of sea-level during the Flandrian stage and the presence of now infilled valleys on the Morecambe Bay area, must have affected the range of tides along the coast. Jardine (1975) has drawn attention to the need to establish former tidal ranges, and has observed that, where the range is great, a Mean Tide Level curve, will be the nearest approximation to a Mean Sea Level curve.

Of all the coastal areas for which there are studies on relative sea-level changes, the Lancashire coast suffers from and records the greatest tidal range, which introduces a further source of error, particularly when pan-European correlations are attempted.

In the Netherlands, Jelgersma records a tidal range for most of the coast within the limits 1.50 to 2.00 metres; only in the south-west of the country, in

Zeeland, does the tidal range increase to 3.00–4.00 metres. Along the Swedish west coast, where Mörner (1969a) has carried out work on sea-level changes, the tidal range is negligible, while in the Baltic, changes in sea-level are caused by meteorological conditions, and the tide is hardly appreciable (Admiralty Tide Tables 1970).

If the geometry of former inlets and bays can be established for different times during the Flandrian stage, then the tidal range may be computed.

Tectonic Warping

It is possible that Flandrian deposits in the tidal flat, lagoonal and perimarine zones have been subject to deformation and downwarping since their formation, because of subsidence along the margins of sedimentary basins, the movement of the basement rocks along fault lines and the evolutions of petroleum and natural gas in West Derby and of rock salt in Amounderness and Cheshire. In addition, the whole area has probably suffered slight subsidence followed by uplift consequent upon deglaciation. The magnitude of the isostatic recovery would increase from south to north, particularly north of Morecambe Bay.

Wray and Cope (1948) show a series of faults in the Keuper and Bunter Sandstones which underlie the Flandrian deposits of coastal West Derby. The Hillhouse and Ince Blundell faults converge on Downholland Moss. In addition, Downholland Moss has been renowned for oil seepages, although Cope (1939) has demonstrated that the occurrences probably derive from the rich diatom flora of the organic marine clays and reedswamp peats. Movement along fault lines may have recommenced during the Devensian or earlier glaciations or since the accumulation of Devensian tills and sands, and Flandrian deposits, recorded to a maximum thickness of 77 metres in Blackpool. In the Netherlands, Jelgersma notes that there is evidence of activity along fault lines during the Old and Middle Pleistocene, but that the Young Pleistocene is probably too short for unequivocal evidence of faulting. However, Sissons (1972) has suggested that the dislocation of the Main Buried Shoreline in the Forth Valley may be associated with the renewed activity along the Abbey Craig fault.

Within the north-eastern Irish Sea are a series of sedimentary basins, the extent and depth of which have been proved by geophysical surveys (Bott, 1968). The results of these surveys show five deep sedimentary basins, of which the East Irish Sea Basin extending from St. Bees Head to the Great Ormes Head includes all the sites, at or near its eastern margins, considered in this present work. The east Irish Sea Basin is probably of Carboniferous and Permo-Triassic age and is filled with sediments of this age and more recent ages to maximum thicknesses of 2.4 to 6.0 km. Bott shows further that the basins are associated with crustal attenuation and that the crust thickens towards Wales and the Lake District. He makes the interesting point that the location of the basins is intimately related to the crustal structure, and that the present coastline, following the margins of the sedimentary basins, is indirectly related to the crustal structure also. The lineation of the present coast, particularly in Cumbria, is intimately related to much older geological episodes, which Bott refers to the Caledonian orogeny. Such a view on the antiquity of the parts of the present coast is sympathetic to that held by Kidson (1971), although in the case of most of the Lancashire coast within the East Irish Sea Basin it is apparent that its lineation is more recent. It is not structurally controlled and is probably the result of events no older than the Upper Pleistocene.

It is possible that slight downwarping is continuing within the basins and that the altitudes of Flandrian marine surfaces have been affected.

From an analysis of tide gauge data, Valentin (1953) has shown that the whole Lancashire coast is undergoing slight downwarping of the order of 0.5 mm/year, whereas Stephens and Synge (1966) show the whole area to be subject to a positive movement of 1 mm/year. Both Wright (1937) and Churchill (1965b) have argued for an elevation of beach deposits in north-west England in response to ice-loading. Wright shows the zero isobase for the Neolithic beach bisecting the Lancashire coast, and Churchill argues that deposits formed at sea-level 6,500 years ago in south-west Lancashire have subsequently been elevated by over 3 metres above present sea-level.

Areas in Amounderness and Cheshire are thought to have undergone subsidence as the result of salt formation, which may have emphasised local features such as Marton Mere, filled since the beginning of the Flandrian stage with a deep, more or less homogeneous gyttja, or the wide-spread subsidence of Over-Wyre. However, there is no evidence from the sediments of Marton Mere that the basin has been affected by overdeepening since its formation as a kettle-hole during the culminating phases of the Devensian glaciation. Furthermore, the height consistencies of the same time/ stratigraphic boundaries in the tidal flat and lagoonal zone from North Wales, through Cheshire to south-west Lancashire and the Fylde, militate against an agrument for anything more than slight downwarping or elevation of Flandrian sediments since their formation. Exact compensation of one by the other seems a highly unlikely mechanism to explain the. height consistencies over a wide area.

Radiocarbon Dating

Shotton (1967) has considered the possible sources of error inherent in obtaining radiocarbon dates, which errors include the contamination of material by both younger and older humous material, the change in the C^{14} content of the atmosphere and the statistical errors inherent in the method. In addition, the radiocarbon dates of dendrochronologically-dated wood samples between 7,300 years BP and the present, demonstrate deviations of the radiocarbon time scale from the absolute time scale of up to 700 years (Suess 1967). Radiocarbon dates during the periods of maximum deviation can be adjusted according to the calibrations of Bristlecone pine, *Pinus aristata*, and Douglas fir, *Sequoia gigantea*, given by Suess (1970a), Damon *et al.* (1970, 1973) and Ralph and Michael (1970). The radiocarbon dates given here have not been adjusted according to this calibration, and all dates are given with a half-life of 5570±30 years. Jelgersma (1961) has considered more particularly the errors that are likely to affect samples dated from the coastal zone, and has made a distinction between dated samples that can be used to specify eustatic movements of sea-level and dated samples that can be used to elucidate the coastal evolution of the Netherlands during the Flandrian stage. She stresses the importance of establishing a chronology independent of the absolute chronology deriving from radiocarbon dates, and argues for series of radiocarbon dates from single cores as an internal check on the dated samples. However, Benzler and Geyh (1969) have shown by dating peat samples from the Kehdinger marsh in the Elbe estuary, that dating a large number of samples from a single core does not permit internal correction; all dates here fall within Flandrian III for which there is no independent means of checking the absolute chronology, such as by correlation of local and regional pollen assemblage zones. In addition, the samples dated ranged from 25 to 40 cm in thickness, and undoubtedly an additional error was thereby introduced.

Geyh (1969) has described in detail the likely errors in dating organic material in tidal flat areas, which errors will affect the establishment of local and regional coastal chronologies during the Flandrian stage.

The material dated in this study comprised monocotyledonous peat which, in many cases, included *Phragmites* rhizomes, which may introduce a systematic dating error as great as 845±210 years, according to the empirical studies of Streif (1972). Microfossil analysis of this material indicated the proximity of marine conditions to the sampling site. Material with a high inorganic content was avoided, because of the possibility of including allochthonous material from estuarine sources which would give an aberrant date. In addition, samples with a high proportion of wood, in the form of roots (*Turfa lignosa*) and branches (*Detritus lignosa*), were avoided, because of the real possibility that the former had penetrated the dated stratum from a higher level, or that the latter had moved downwards through an accumulating detrital deposit with a high ground-water table. Godwin *et al.* (1935) and Kaye and Barghoorn (1964) have noted that tree trunks and branches tend to suffer settlement within a detrital peat and to suffer auto-compaction, so that branches of different ages can lie adjacent to each other in a particular stratum, and downward movement is halted when there is a change in lithology: the latter situation can be seen at West Hartlepool while the former is abundantly clear in parts of the *Salix*-dominated fen of Lough Cranstal, Isle of Man.

There is always the possibility of contamination by rootstocks which may or may not be evident in fossil material: Benzler and Geyh (op. cit., p. 362) note that 30 per cent of the rootstock comprises root hairs that cannot be detected, and these can be a serious source of contamination. Further, we do not agree

Table 1. Radiocarbon Dates

NO	LABORATORY CODE	YEARS BP	HEIGHT OD METRES	MATERIAL	SITE NAME AND NUMBER	NATIONAL GRID REFERENCE
1	Birm 139	9195±155	-16.37	Peat	Heysham Head M1.18	SD 39365989
2	Birm 140	8925±200	-16.04	Peat	Heysham Head M2.13	SD 39335988
3	Birm 141	9270±200	-17.60	Peat	Heysham Head M3.17	SD 39325990
4	Birm 230	12320±155	-0.97	Gyttja	Rossall Beach	SD 31114383
5	Q 85	5277±120	+4.88	Phragmites peat	Helsington Moss	nk
6	Q 88	4616±112	+5.18	Peat	High Foulshaw Moss	nk
7	Q 256	5734±129	+2.93	Fen peat	Silverdale Moss	nk
8	Q 261	5865±115	+3.85	Alnus wood	Silverdale Moss	nk
9	Q 620	3695±110	+3.05	Wood	Reeds Lane, Moreton	nk
10	Hv -	170±65	+2.53	Scrobicularia plana complete valves	Formby Foreshore	SD 27000858
11	Hv 2679	4545±90	+3.14	Phragmites peat	Alt mouth	SD 29500290
12	Hv 2680	no date	+0.13	Clay gyttja	Downholland Moss - 15	SD 32020838
13	Hv 2680A	6750±175	+0.15	Clay gyttja	Downholland Moss - 15	SD 32020838
14	Hv 2681	4325±345	+0.30	Phragmites peat	Downholland Moss - 15	SD 32020838
15	Hv 2682	4600±430	+1.06	Phragmites peat	Downholland Moss - 15	SD 32020838
16	Hv 2683	5565±205	+1.27	Phragmites peat	Downholland Moss - 15	SD 32020838
17	Hv 2684	4045±395	+1.86	Phragmites peat	Downholland Moss - 15	SD 32020838
18	Hv 2685	5250±385	+1.29	Phragmites peat	Helsby Marsh - 1	SJ 48847614
19	Hv 2686	5470±155	+0.73	Wood detrital peat	Helsby Marsh - 1	SJ 48847614
20	Hv 2916	2270±65	+4.22	Gyttja Phragmites	Lytham Hall Park - 8	SD 34902858
21	Hv 2917	3090±135	+3.70	Phragmites peat	Lytham Hall Park - 8	SD 34902858
22	Hv 2918	3150±150	+3.51	Phragmites peat	Lytham Hall Park - 8	SD 34902858
23	Hv 2919	4960±210	+3.39	Phragmites peat	Lytham Hall Park - 4	SD 34712879
24	Hv 2920	4190±150	+4.49	Woody detrital peat	Heysham Moss - 1	SD 42396083
25	Hv 3052	4900±450	+4.80	Phragmites peat	Lousanna	SD 41604486
26	Hv 3356	8685±175	-15.48	Peat	Morecambe Bay C8	SD 33507465
27	Hv 3357	4695±110	+0.24	Phragmites peat	Downholland Moss - 15	SD 32020838
28	Hv 3358	6050±65	+1.06	Phragmites peat	Downholland Moss - 15	SD 32020838
29	Hv 3360	7725±95	-15.84	Peat	Morecambe Bay C7	SD 32207791
30	Hv 3361	8740±65	-16.54	Peat	Morecambe Bay A5	SD 32937636
31	Hv 3362	7995±80	-11.16	Peat	Morecambe Bay B1	SD 32087190
32	Hv 3460	5015±100	+4.98	Peat	Arnside Moss	SD 46727895
33	Hv 3461	1545±35	+5.65	Peat	Arnside Moss	SD 46727895
34	Hv 3462	8330±125	-16.03	Peat	Morecambe Bay C6	SD 44866646
35	Hv 3840	4760±45	+5.29	Phragmites peat	Duddon Estuary - 11	SD 22058477

36	Hv 3841	4960±50	+4.06	Woody detrital peat	Duddon Estuary - 16	SD 22258520
37	HV 3842	2820±55	+6.66	Gyttja Phragmites	Annas Mouth	SD 07688841
38	Hv 3843	7160±75	+0.91	Gyttja Betula	Bootle Beach	SD 07959090
39	Hv 3844	5435±105	+3.72	Phragmites peat	Ellerside Moss	SD 34958024
40	Hv 3845	5005±65	+3.09	Phragmites peat	Lytham Common - 1A	SD 33092947
41	Hv 3846	830±50	+6.52	Monocot. peat	Lytham Hall Park - 6	SD 34902858
42	Hv 3847	5615±45	+1.59	Phragmites peat	Downholland Moss - 16A	SD 32380839
43	Hv 3933	4800±75	+2.24	Woody detrital peat	Peel - 2A	SD 32123578
44	Hv 3934	6535±110	-0.83	Woody detrital peat	Peel - 2A	SD 32123578
45	Hv 3935	6760±95	-0.14	Phragmites peat	Downholland Moss - 11A	SD 33650819
46	Hv 3936	6980±55	-0.38	Phragmites gyttja	Downholland Moss - 11A	SD 33650819
47	Hv 4124	5945±50	+0.88	Monocot. peat Phragmites	Nancy's Bay - 10	SD 37962863
48	Hv 4125	7605±85	-2.33	Monocot. peat	Nancy's Bay - 10	SD 37962863
49	Hv 4126	6885±80	-1.17	Monocot. peat	Nancy's Bay - 10	SD 37962863
50	Hv 4127	6025±85	+0.46	Phragmites peat	Nancy's Bay - 10	SD 37962863
51	Hv 4128	5775±85	+1.93	Phragmites peat	Nancy's Bay - 6	SD 38372799
52	Hv 4129	5950±85	+1.52	Phragmites peat	Nancy's Bay - 6	SD 38372799
53	Hv 4130	6245±115	+1.13	Phragmites peat	Nancy's Bay - 6	SD 38372799
54	Hv 4131	6290±85	+0.97	Phragmites gyttja	Nancy's Bay - 6	SD 38372799
55	Hv 4343	8390±105	-11.14	Woody detrital peat	Starr Hills	SD 33522747
56	Hv 4344	4895±95	+2.87	Phragmites gyttja	Heyhouses Lane LC-2A	SD 33942946
57	Hv 4345	7820±60	-9.65	Gyttja	Heyhouses Lane LC-2A	SD 33942946
58	Hv 4346	8575±105	-9.75	Gyttja	Heyhouses Lane LC-2A	SD 33942946
59	Hv 4347	4830±140	+2.95	Phragmites gyttja	Moss Farm - 6	SD 44544837
60	Hv 4348	4725±65	+2.43	Woody detrital peat	Rhyl Beach - 1	SJ 03108261
61	Hv 4417	805±70	+5.63	Monocot. peat	Lytham Common - 1A	SD 33092947
62	Hv 4418	4845±100	+8.70	Gyttja	Pelutho - 1	NY 12124923
63	Hv 4705	4090±170	+3.42	Monocot. peat	Downholland Moss - 3C	SD 31270817
64	Hv 4706	6950±175	-2.46	Monocot. peat	Nancy's Bay - 10	SD 37962863
65	Hv 4707	5880±180	-1.22	Monocot. peat with Phragmites	Nancy's Bay - 10	SD 37962863
66	Hv 4708	1370±85	+4.60	Monocot. peat	Rossall Road, Ansdell-1	SD 34902775
67	Hv 4709	2335±120	+5.08	Woody detrital peat	Formby Foreshore 2	SD 26950642
68	Hv 5215	1795±240	+4.04	Monocot. peat	Ansdell-2	SD 34822754
69	Hv 5294	6250±55	+1.39	Gyttja	Nancy's Bay-6	SD 38372799

with Geyh that a peat-horizon sealed by marine sediments gives any immunity at all to the peat, which can be contaminated with more recent roots penetrating both the clay and the peat strata. Kaye and Barghoorn (op. cit.) note that the rootlets of *Spartina* penetrate 38 cm below the rhizomes, and in open excavations in Lytham Hall Park, I have observed the rhizomes of *Phragmites* penetrating tree trunks and passing through thin estuarine clay strata from one peat layer to another.

Notwithstanding the very great care exercised in the acquisition of samples for radiocarbon dating, results of pollen analyses carried out on these samples and correlation of the local pollen assemblage zones with the regional pollen assemblage zone have shown that some of the radiocarbon dates are in error and could not be used in either the sea-level curve graph or the scheme of Flandrian coastal sequences. The dates are considered together with their respective stratigraphic and pollen analytical investigations in Chapter 2, and all the radiocarbon dates are displayed in Table 1.

Sites Investigated

The Lancashire and Cheshire coasts, with their numerous deep estuaries and extensive flats are noted . . . for their submerged forests, sometimes seen on the foreshore between tide marks, sometimes laid open in extensive dock or harbour works. . . . The estuaries of the Ribble, Mersey and Dee tell a similar story, for on their shores and under their marshes are found some of the most extensive submerged land-surfaces now traceable in Britain . . . We have in these strongly marked alternations of peat and warp an ideal series of deposits for the study of successive stages. In them the geologist should be able to study ancient changes of sea-level under such favourable conditions as to leave no doubt as to the reality and exact amount of these changes. (Clement Reid 1913.)

The early promise that estuarine and open coast sites in north-west England held out for a study of sea-level changes has been fulfilled by recent work on the submerged forests and layers of peat and clay recorded along these coasts (Tooley 1969, 1970, 1974, 1976b, 1977a, b, Huddart *et al*. 1977). Altogether thirty-six sites have been examined and several hundred borings made along the coastal margins of the East Irish Sea basin (Fig. 2). Sites flanking the Ribble estuary have been selected for detailed consideration here. They have been chosen for two reasons: first, to demonstrate the methodology that has proved effective in resolving problems associated with sea-level studies in humid temperate middle latitudes; second, to present critical data for the sea-level curve and the sequence of marine transgressions, for which Lytham, on the north side of the Ribble estuary, is the type area.

LANCASHIRE: WEST DERBY

In the western part of West Derby (Fig. 2), solid rock approaches the surface and forms surface features only locally. The rocks are Triassic in age and comprise Keuper and Bunter Sandstones: Keuper Sandstone is exposed in Altcar Quarry at Hillhouse (Figs. 2 and 14). Elsewhere Triassic rocks have been proved in borings. A comprehensive review of the extent and structure of these rocks is given by Wray and Cope (1948). Throughout the area, solid is obscured by till and by more recent deposits of varying thicknesses.

The altitude of the sub-drift surface varies from + 6.0 metres O.D. north of Downholland Cross (SD 374075) to −26.8 metres O.D. on Downholland Moss (SD 318079). At the new River Alt pumping station, the lowest altitude of the subdrift surface was recorded at −11.44 metres O.D. Howell (1965) gives comprehensive lists of drift thickness in south-west Lancashire: west of Halsall (SD 368108) the drift is 2.4 metres thick, whereas on Downholland Moss it is 30.5 metres thick and exceeds 32.3 metres in Southport (SD 349163). Howell derives these figures by deducting the altitude of the subdrift surface from the known or interpolated ground altitudes: this derivation therefore includes sediments of post-Devensian age in many cases. Locally, Flandrian deposits exceed 18 metres in thickness (SD 29440762), and thus the error in the thickness of till units calculated can be of this order. At the River Alt pumping station, the thickness of the till units never exceeds 6 metres.

In the vicinity of Downholland, Howell (1973) shows a linear funnel-shaped valley in the sub-drift surface from sea-level to −100 ft O.D.: this depression may be a section of the lower part of the proto-Rainford river valley after its confluence with the proto-Alt. The morphology of the valley has been maintained until the middle of the Flandrian stage. West of the fifteen metre contour, the till surface is obscured by more recent deposits of Late Devensian and Flandrian age, the stratigraphy of which has been shown in cross-sections published by Reade in 1871 (Fig. 14).

The till surface is overlaid by head or by blown sand, known as Shirdley Hill Sand, after its type locality (Figs. 2 and 14). The sand is subdivided into distinct units by intercalated beds of unhumified peat or organic sand, which are well developed in shallow basins and depressions in the till. Occasionally the sand exceeds 5 metres in thickness, and has been reported in borings from some way below Ordnance Datum to 120 metres above it (Wray and Cope 1948). West of Haskayne (SD 37920693) the sand forms parabolic dunes of modest altitude, which dunes have been mapped by Gresswell (1953, Fig. 106, p. 27). Gresswell, following de Rance (1869), has always maintained (1953, pp. 46-8, and 1957, pp. 68-70) that the Shirdley Hill Sands are coastal in origin. They began to accumulate during the late Atlantic and early Sub-Boreal periods after a fall in relative sea-level of the order of 12 metres, followed by a rise in sea-level up to the altitude of the *Hillhouse coastline*. If this was so, then any lagoonal, terrestrial or dune slack peat in the Shirdley Hill Sand formation would furnish evidence, not only of the vegetation composition and relative age of the peat, but also the altitude of relative sea-level at the time of inundation.

Preliminary results from palaeobotanical investigations of the intercalated biogenic deposits and mechanical analyses of the sands (Tooley and Kear 1977) indicate that the Shirdley Hill Sands are Late Devensian in age, and this is supported by the radiocarbon dates. Their wide extent suggests analogies with the cover-sands of Europe—a conclusion reached by Godwin in 1959.

West of the +7.0 metre contour, the till surface and Shirdley Hill Sands pass under a suite of clays and silts, intercalated with biogenic deposits, or directly under peat. These sedimentary units are best developed on Downholland Moss, although they can be found throughout the low-lying ground from the River Douglas to the River Alt, between the rising till to the east and the blown sand to the west.

The upper peat, which reaches five metres in thickness locally (Hall 1954-5) has been subject to wastage from cutting and drainage operations, and little, if any, of the original raised moss surface remains. Locally, coverts stand higher than the surrounding farmland, and carry a wind-streamed scrub of *Betula*, *Acer* and *Salix* and a dense shrub layer of *Rhododendron ponticum* Linné, *Rubus fruticosus* agg. and *Crataegus monogyna* Jacq. Where recent burning has occurred, there is a flush of *Chamaenerion angus-*

tifolium (L.) Scop. and *Pteridium aquilinum* (L.) Kuhn. On Downholland Moss, the surface of the coverts lies less than a metre above the reclaimed farmland: at DM-16 the covert is at +3.2 metres O.D. compared with +2.4 metres O.D. in adjacent fields. In Shire Hall Meadow on Downholland Moss (SD 32000830), the thickness of the upper peat was recorded as 1.5 metres in 1954 (Hall, loc. cit.), but by 1968 it was only 0.5 metres. On Martin Mere, Heath Covert (SD 390162) rises 2 metres above the surrounding farmland.

To the seaward of the mosses, a belt of sand 0.5 to 4.8 km wide and attaining over 30 metres in altitude in the Ainsdale and Birkdale Hills, has drifted over and interdigitates the peat on the mosslands. The over-burden of sand increases in thickness westward. Thus, at DM-4 (SD 31560817) there is a sandy soil 50 cm thick obscuring the clay/peat series there, while at DM-2 (SD 30940813) there is an over-burden of 120 cm of sand, and at Long Lane, Formby (SD 29440762) it is 442 cm thick. The sand is articulated by organic layers, similar to those reported from the Fylde.

Downholland Moss

Downholland Moss (Figs. 2 and 5) is a low-lying arable area to the east of Formby. It is bounded in the north by Downholland Brook and by New Cut in the south, beyond which is Altcar Moss. It is part of an inter-connected series of mosses and freshwater lakes which extended from the margins of the former Martin Mere in the north to Altcar Moss in the south.

The lowest part of the moss is around sampling site DM-8, where the ground altitude is +1.82 metres O.D. To the east, the surface rises to +3.20 metres, as the till sub-surface rises towards Haskayne. The ground surface also rises westwards to a maximum altitude of +5.15 metres at DM-2, as the over-burden of blown sand thickens, and reaches its furthest eastern penetration on to the Moss between DM-4 and DM-5.

The stratigraphy of the Moss has been described by Binney and Talbot (1843), by Wray and Cope (1948) and by Hall (1954-5). Binney and Talbot described the stratigraphic succession in the area bounded by Barton Brook and the Fleam Brook:

1. Black, decomposed peat, having scarcely any vegetable structure in it; the lower part hard and of a pitchy nature. 0 to 0.67 feet

2. Black peat, very moist, full of petroleum, but showing little trace of vegetable structure, except in the lower parts of it, called in the neighbourhood 'Pipy Moss', from the circumstances of its containing the compressed stems of reeds and other plants. 0.67 to 3.83 feet

3. Soft, blue silty clay. 3.83 to 7.33 feet

4. Dry, mouldy peat containing seeds resembling mustard seed, of a brown colour when fresh obtained, but it soon turned black on exposure to the atmosphere. The lower part of this bed, like No. 2, was reedy, but it possessed no empyreumatic smell. 7.33 to 11.33 feet

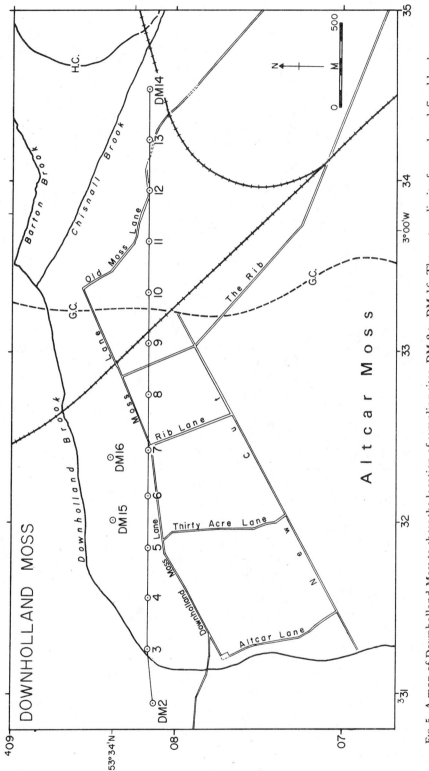

Fig. 5. A map of Downholland Moss to show the location of sampling sites, DM-2 to DM-16. The eastern limit of grey clays defined by de Rance (1869) is shown by the pecked line marked G.C., while the Hillhouse coastline (H.C.) described by Gresswell (1953) is shown by a continuous line, where it is a marked landscape feature.

5. Blue, silty clay.	11.33 to 11.83 feet
6. Blue, sandy clay penetrated.	11.83 to 15.33 feet

Fifteen borings were put down across Downholland Moss and the stratigraphy at each sampling site recorded. The location of each site is shown on Fig. 5 and the stratigraphy drawn up on Figs. 6 and 7.

At three sampling sites (DM -11, -12 and -13) the sub-surface of the till was encountered beneath a veneer of Shirdley Hill Sand, at depths from the surface of 430, 223 and 145 cm respectively. South of DM-5, Wray and Cope (1948, p. 46) record the till sub-surface at 19.8 metres below the surface (−14.33 metres O.D.). Thus between DM-11 and DM-5, there is an imperceptible gradient of the till sub-surface of 1:166.

Deposits of Flandrian age increase in thickness westwards, however. On Downholland Moss, a maximum of 4 metres has been proved at DM-11 and 19 metres south of DM-5 (Wray and Cope, loc. cit.). 2.5 km west of DM-5, at Long Lane, Formby, there are 18 metres of Flandrian deposits, but the altitude of the till sub-surface has risen to about −10.90 metres O.D.

On Downholland Moss, four sites DM-7, DM-11, DM-15 and DM-16 were chosen for detailed analysis, because of the variety of stratigraphic types that they displayed and three of these sites, DM-11, DM-15 and DM-16, are described here. Additional data can be found elsewhere (Tooley 1969, 1977a).

Downholland Moss — 11, DM-11, Thomas Whalley's New Moss — 1

This site was located towards the eastern margin of Downholland Moss, where the till sub-surface was rising (see Fig. 6), but still just within the tidal flat and lagoonal zone. This site and an adjacent one (DM-11A) were sampled with a Hiller and with a Russian sampler, and samples collected for pollen and radiometric analyses. The site had two advantages over other sites on Downholland Moss: the single clay layer, Stratum 3, overlaid a coarse organic sand, which in turn passed down into till, and it was unlikely that consolidation would have affected greatly the altitude of Stratum 3; the second advantage was that the long uninterrupted biogenic accumulation of DM-11 above Stratum 3, allowed the development of pollen assemblages which were more representative of a larger area than those from other sampling sites on the moss, where biogenic deposition was interrupted by sedimentation of marine clays, silts and sands.

From DM-11, the following stratigraphic succession was recorded:

Stratum	Height O.D. metres	Depth	Description
12	+2.77 to +2.52	000 to 025	Str. conf. Peaty soil.
11	+2.52 to +2.22	025 to 055	Sh [4] 2, Dl.2, nig. 4, strf. 0, elas. 0, sicc. 3, lim. sup. 0. An oxidised peat, becoming woody, compact, hard and difficult to penetrate towards the base.

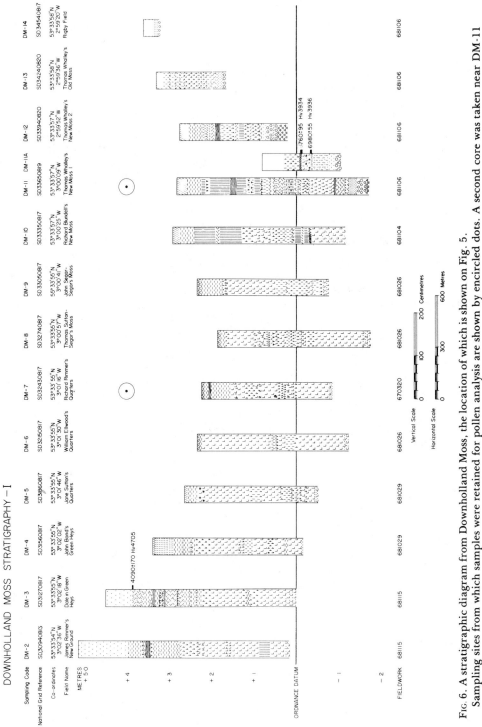

FIG. 6. A stratigraphic diagram from Downholland Moss, the location of which is shown on Fig. 5. Sampling sites from which samples were retained for pollen analysis are shown by encircled dots. A second core was taken near DM-11 for radiocarbon dating and the stratigraphy is shown adjacent to this core as DM-11A.

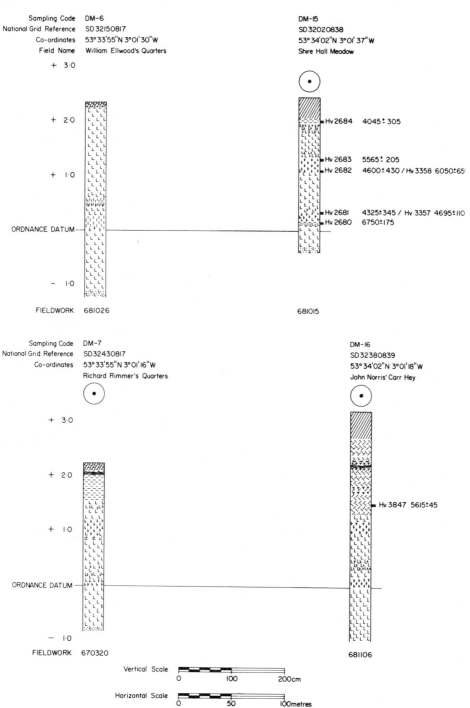

DOWNHOLLAND MOSS   STRATIGRAPHY — II

Sampling Code   DM-6
National Grid Reference   SD 32150817
Co-ordinates   53° 33'55"N 3°01'30"W
Field Name   William Ellwood's Quarters

+ 3·0

+ 2·0

+ 1·0

ORDNANCE DATUM

— 1·0

FIELDWORK   681026

DM-15
SD 32020838
53° 34'02"N 3°01'37"W
Shire Hall Meadow

Hv 2684   4045 ± 305

Hv 2683   5565 ± 205
Hv 2682   4600 ± 430 / Hv 3358 6050 ± 65

Hv 2681   4325 ± 345 / Hv 3357 4695 ± 110
Hv 2680   6750 ± 175

681015

Sampling Code   DM-7
National Grid Reference   SD 32430817
Co-ordinates   53° 33'55"N 3°01'16"W
Richard Rimmer's Quarters

+ 3·0

+ 2·0

+ 1·0

ORDNANCE DATUM

— 1·0

FIELDWORK   670320

DM-16
SD 32380839
53° 34'02"N 3°01'18"W
John Norris' Carr Hey

Hv 3847 5615 ± 45

681106

Vertical Scale
0   100   200cm

Horizontal Scale
0   50   100metres

FIG. 7. An additional stratigraphic diagram from Downholland Moss, showing the relationship of the two sampling sites from which samples for radiocarbon dating were taken. Sites for which there are pollen diagrams are shown by encircled dots.

10	+2.22 to +1.52	055 to 125	Th ^{4} 2, Th ^{1} *(Meny.)*+, Ld ^{4} 2, nig. 3, strf. 1, elas. 0, sicc. 3, lim. sup. 0. More or less completely humified monocot. peat, greasy, muddy and fine-textured in places, with *Menyanthes* seeds at 69 cm.
9	+1.52 to +1.39	125 to 138	Tb ^{2} 4, nig. 2, strf. 0, elas. 2, sicc. 1, lim. sup. 0. Partly humified, *Sphagnum* peat.
8	+1.39 to +0.84	138 to 193	Ld ^{4} 2, Th ^{4} 1, Th ^{1} *(Meny.)*+, Dl.1, nig. 3, strf. 1, elas. 0, sicc. 2, lim. sup. 0. Gyttja with woody and monocot. fragments throughout, including *Menyanthes* seeds at 176, 165 to 170 cm.
7	+0.84 to +0.22	193 to 255	Th ^{3} 2, Th ^{1} *(Meny.)*1, Dl.2, anth+, nig. 3, strf. 1, elas. 1, sicc. 2, lim. sup. 0. Partly humified, monocot. peat with woody detrital remains including *Alnus* at 214, 225 and 235 to 245, *Menyanthes* seeds at 193 cm., and charcoal at 194 to 195 and at 252 cm.
6	+0.22 to +0.04	255 to 273	Ld ^{4} 3, Th ^{3} 1, Th ^{2} *(Phra.)*+, nig. 3, strf. 1, elas. 0, sicc. 2, lim. sup. 0. Gyttja with some monocot. fragments throughout, including *Phragmites*.
5	+0.04 to -0.11	273 to 288	Dl.4, nig. 4, strf. 2, elas. 0, sicc. 2, lim. sup. 0. Woody detrital peat with *Alnus* branches.
4	-0.11 to -0.36	288 to 313	Th ^{3} 2, Th ^{1} *(Phra.)*2, Dl+, nig. 2, strf. 2, elas. 1, sicc. 2, lim. sup. 0. Monocot. peat, rich in *Phragmites* with a little *Alnus* wood at 307 cm.
3	-0.36 to -0.87	313 to 364	As3, Sh ^{4} 1, Th ^{1} *(Phra.)*+, nig. 1, strf. 1, elas. 0, sicc. 2, lim. sup. 0. Tenacious, blue clay, with organic material, including *Phragmites* richly toward the top and base of the stratum, and within the stratum at 325, 337 and 345 cm.
2	-0.87 to -1.23	364 to 400	Ga2, Sh ^{4} 2, Th ^{1} *(Phra.)*+, Dl+, nig. 4, strf. 0, elas. 0, sicc. 2, lim. sup. 0. Black organic sand, with *Phragmites* at 387 cm. and *Alnus* wood occasionally, e.g. at 393 cm.
1	-1.23 to -1.73	400 to 450	Ga1, Gs2, Gg (maj.)1, nig. 0̄, strf. 0, elas. 0, sicc. 2, lim. sup. 1. Sand, becoming coarse with stones and impenetrable towards the base. Sand white.

The stratigraphic succession at both DM-11 and DM-11a, from where samples for radiocarbon dating were obtained, is shown diagrammatically on Fig. 8, together with a pollen diagram from DM-11 (Fig. 8). Five Local Pollen Assemblages Zones have been recognised:

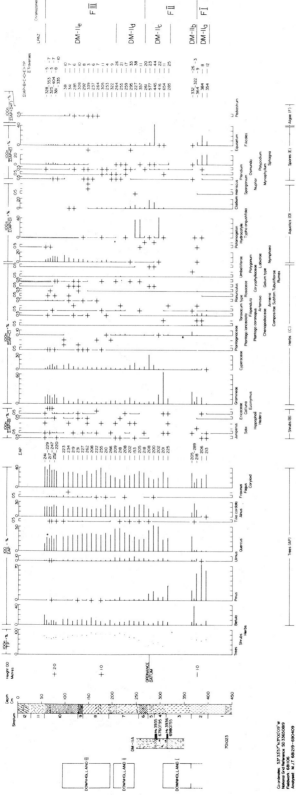

Fig. 8. Pollen diagram from Downholland Moss-11. The percentage frequency of each taxon is based on the calculating formula given for each of the six groups recognised AP, Tree taxa: B, Shrub taxa: C, Herb taxa: D, Aquatic taxa: E, Pteridophyte and Bryophyte taxa: F, Algae. The stratigraphy and radiocarbon dates from DM-11A are also shown. Three of the transgressions recognised on Downholland Moss are shown on the basis of the mean heights of the transgressive and regressive phases: the transgressions correspond to Lytham III (Downholland I), Lytham IV (Downholland II) and Lytham VI (Downholland III).

LPAZ	*Depth cm.*	*Zone Description*

DM-11a   395
          to
          375

The AP component accounts for over 50%ΣTP
throughout the zone, and of this the highest
proportion is contributed by *Pinus* which attains
a maximum frequency of 74%ΣAP at 385 cm.
Ulmus values are discontinuous and *Quercus*
values are low but continuous:there is the
beginning of a continuous *Alnus* record from
385 cm. Groups AP, C and E dominate the
assemblage. There are low, continuous frequencies
of Tubuliflorae and *Filipendula* and moderate,
declining frequencies of Sphagna spores.

DM-11b   375
          to
          365

The AP Group maintains its preponderance over
other groups, accounting for over 60%ΣTP and
rising to 90% at 370 cm. The behaviour of tree
taxa characterises the zone: there is an
isolated peak of *Betula* at 370 cm. of 50%ΣAP,
Pinus values collapse from 55% at 375 to 4% at
370 cm., *Quercus* values rise suddenly from 3%ΣAP
at 370 cm. to 31% at 365 cm. *Alnus* also rises
to a peak value of 28%ΣAP at 370 cm., from which
it then declines. Group C taxa are the only
other group well represented: Grass values rise
through the zone, and there is the appearance
of the Chenopodiaceae.

DM-11c   315
          to
          265

The proportion of the AP group falls suddenly at
the base of the zone, sustains frequencies less
than 45%ΣTP, and rises again at the end of the
zone: this recession is in face of the
increasing proportions of Groups C, D and E at
different levels within the zone. *Quercus* rises
to maximum values of 53%ΣAP at 285 cm. from which
level it then declines: the rise in frequency is
at the expense of both *Alnus, Pinus* and Coryloid
which behave in complement. *Ulmus* frequencies
are low throughout, but tend to fall through the
first part of the zone and rise in the latter
part: *Ulmus* accounts for 6%ΣAP at 305 cm.,
falls to 1% at 285 and rises to 6% again at
265 cm., which rise is sustained into zone d.
Grass values attain an isolated peak towards the
base of the zone, from which level they then
decline, whilst the Cyperaceae rise slightly.
Of Group D taxa there is an isolated peak of
Typha angustifolia of 42% (AP+D) at 295 cm.
succeeded by a modest rise in the frequency of
Cladium mariscus to 18% (AP+D) at 275cm.

LPAZ	*Depth cm.*	*Zone Description*

DM-11d   265
          to
          205

The AP component continues to be the most
important single group: *Quercus* values, after
falling initially, rise steadily throughout
the zone. There are slight changes in the
moderate frequencies of both *Alnus* and Coryloid.
Ulmus maintains low, continuous frequencies
between 5 and 11% AP and the d/e zone boundary
is drawn at 205 cm., where the *Ulmus* values

fall from 8% at 205 cm. to 2% at 195 cm. and thereafter never exceed 2% AP. Taxa in Group D behave in a similar fashion to the pattern recorded in zone c: *Typha angustifolia* once again rises to peak values of 38% (AP+D) at 245 cm., whilst *Cladium mariscus* rises slightly to 8% at 245 cm. and then declines.

DM-11e   205
          to
          55

The feature of this zone is the remarkable uniformity of the proportions of the AP taxa frequencies, which are dominated by *Quercus*, *Alnus*' and Coryloid throughout. *Pinus* becomes discontinuous at 175 cm. and *Ulmus* at 70 cm. The number and frequency of open habitat taxa increase throughout the zone: in particular *Plantago lanceolata* is first recorded at 125 cm. and there are low frequencies or isolated occurrences of *Artemisia*, *Galium*-type and *Rumex*. Of Group D taxa, *Cladium* maintains low frequencies and the record becomes discontinuous above 145 cm., whilst the zone opens with a continuous *Nymphaea* curve, and frequency of the taxon increases throughout the zone to 14% (AP+D) at 95 cm.

Two samples were taken from the adjacent sampling site at DM-11A with a Russian-type sampler for radiocarbon dating. The sample from Stratum 3 at 313 to 316 cm correlated with Stratum 2 at DM-11 gave a date of 6980±55 and the sample from Stratum 5 at 289 to 293 cm correlated with Stratum 4 at DM-11 gave a date of 6760±95 (Table 1, Nos. 46 and 45). On the basis of these dates, it is possible to assign Strata 3, 4 and 5 in DM-11A to Chronozone II. These can be related with some confidence to the pollen diagram at DM-11, not only on the basis of the similar succession of lithostratigraphic units at the two sampling sites but also by comparing the behaviour of the tree taxa in Stratum 2 at DM-11 with those at Red Moss, Horwich. At Red Moss, Hibbert *et al.* (1971) show for the chronozone boundary zone FId/FII, a collapse in pine frequencies from a maximum of c. 45% ΣAP in the middle of zone FId to c. 28 per cent just below the chronozone boundary and about 6 per cent just above it. In complement, *Alnus* frequencies rise suddenly immediately below the boundary. Although the actual frequencies differ, as would be expected when comparing an inland site with one in the tidal flat and lagoonal zone, the behaviour of these taxa is identical. The radiocarbon date at the FId/FII boundary at Red Moss, is given as 7108±120 BP (Q.916) which compares with the date of the lithostratigraphic boundary at DM-11 of 6980±55 BP (Hv.3936).

A FII/FIII chronozone boundary is drawn tentatively at 205 cm, corresponding to the LPAZ boundary DM-11d/e on the basis of the fall in the frequency of *Ulmus* (See Fig. 4).

Downholland Moss—15, DM-15, Shire Hall Meadow

The sampling site at DM-15 in Shire Hall Meadow was the second site recorded on the Moss. It was chosen because Hall (1954–5) had indicated the thickest over-burden of peat at this site, but the sample was taken before the complete stratigraphic survey of the Moss had been completed, and other sites would have been preferred in the light of the results of this survey (Figs. 6 and 7). In 1968, the Soil Survey of England and Wales permitted their Proline Corer to be used on the Moss to take a continuous core at DM-15, and the results described below derive from analyses carried out on this core. The 15 cm diameter core was obtained in metre-long sections from the south-east corner of Shire Hall Meadow, some ten metres away from the intersection of two drainage gutters. Samples were obtained for pollen and diatom analyses from this core, the stratigraphy was also recorded from the core, after which bulk samples were cut for radiometric and particle size analyses. It had been intended to use this site as the type locality for the Flandrian marine sequences in north-west England, but there were difficulties with the radiometric analyses, and since 1968 the fuller Flandrian marine sequences in Lytham have been proved, and the latter area has been adopted as the type locality for north-west England.

At DM-15, the following stratigraphic succession was recorded in 1976 (it differs from the succession shown in Figs. 7 and 9, which was recorded in 1968):

Stratum	*Height O.D. metres*	*Depth*	*Description*
25	+2.43 to +2.37	000 to 006	Disturbed.
24	+2.37 to +2.24	006 to 019	Gs1, Ld ^{3} 3, Dl.+, Gg (min.)+, nig. 3, sicc. 3, strf. 0, elas.+, struct. crumbly. Sandy gyttja with small wood fragments and grit.
23	+2.24 to +2.11	019 to 032	Ld ^{3} 4, Gs+ Th ^{2} +. nig. 3+, sicc. 3, strf. 0, elas.+, struct. crumbly. Black, crumbly gyttja with discrete sandy inclusions and very occasional monocot. rootlets.
22	+2.11 to +2.04	032 to 039	Ld ^{2} 3, Th ^{2} 1. nig. 2½, sicc. 3, strf. 1, elas. 1, struct. slightly laminated. Very dark brown gyttja, slightly laminated with monocot. roots. No sand.
21	+2.04 to +1.94	039 to 049	Ld ^{2} 3, Th ^{2} 1, Th *(Phra.)* ^{2} +, Dh +. nig. 2½, sicc. 3, strf. 2+, elas. 2, struct. laminated. Dark brown laminated gyttja with monocot. roots throughout and *Phragmites* rhizomes, occasional monocot. and dicot. leaves.
20	+1.94 to +1.89	049 to 054	Ld ^{2} 2, Th *(Phra.)* ^{2} 1, Th ^{2} 1, Dh + Ga+. nig. 2, sicc. 3, strf. 3, elas. 2, struct. well-laminated. Gyttja, well-laminated with *Phragmites* rhizomes and monocot. leaf fragments. Very slight sand fraction.

19	+1.89 to +1.81	054 to 062	Ld 2, As 1, Th[1]+,Th *(Phra.)*[1] l. nig. 2, sicc. 3, strf. 3+, elas. 2, struct. laminated. Green-grey gyttja with discrete iron staining and monocots. throughout. *Nymphaea* seed at 57 cm. Well-laminated.
18	+1.81 to +1.50	062 to 093	As3, Ag1, Ld [4] +, Lf+, Dh+. nig. 2, sicc. 3, strf. 1, elas. 0, struct. plastic. lim. sup. 0. Silty clay with very slight gyttja content, discrete iron partings with sand lenses, horizontally laminated. Penetrated by occasional rootlets, channels iron stained.
17	+1.50 to +1.32	093 to 111	As 2, Ag1, Ld [4] 1, Lf+, Gs+, T [2] +. nig. 2, sicc. 3, strf. 1, elas. 0, struct. plastic, lim. sup. 0. Silty clay gyttja, penetrated by woody roots and v. occasional iron-stained sandy partings.
16	+1.32 to +1.25	111 to 118	As1, Ag1, Ld [4] 1, Lf1, Tl[2] +. nig. 2+. sicc.3, strf. 1, elas. 0, struct. plastic, lim. sup. 0. Silty clay gyttja rich in woody partings and stained throughout with iron, red-brown and yellow in discrete partings.
15	+1.25 to +1.05	118 to 138	Ld [2] 3, Th *(Phra.)*[2] 1, Gs+, Ag+, Th [2] +. nig. 3, sicc. 3, strf. 2, elas. 2, struct. laminated, lim. sup. 0. Slightly laminated dark brown gyttja with occasional sand and silt partings in the laminae. Monocot. rootlets and *Phragmites* rhizomes throughout.
14	+1.05 to +1.03	138 to 140	Ld [2] 2, As1, Lf+, Th [2] 1. nig. 2+, sicc. 3, strf. 1, elas. 2, struct. slightly laminated. lim. sup. 0. Clay gyttja with iron staining in root channels. Slightly laminated.
13	+1.03 to +0.92	140 to 151	Ld [4] 1, As3, Gs+, Th [2] +, Lf+. nig. 2, sicc. 3, strf. 0, elas.0, struct. homogeneous, lim.sup. 0. Clay with a gyttja content. Discrete sand partings and v. occasional iron staining along root channels.
12	+0.92 to +0.71	151 to 172	Ag2, Ga2, Th [3] +, Lf+. nig. 2, sicc. 2, strf. 3, elas. 0. struct. laminated, lim.sup. 0. Laminated sandy silt with very occasional root partings, channels iron-stained. Some stress structures.
11	+0.71 to +0.62	172 to 181	Ag2, As2, Ld [4] +, Lf+, Ga+. nig. 2+, sicc. 2. strf. 1. elas. 0, struct. homogeneous. lim. sup. 0. Silty clay with a slight gyttja content and sand laminae very occasionally.
10	+0.62 to +0.54	181 to 189	Ld [3] 2, As1, Ag1, Ld [4] +. nig. 2[+], sicc. 2, strf. 0, elas. 0, struct. homogeneous, lim. sup. 0. Silty clay gyttja with discrete gyttja partings.
9	+0.54 to +0.44	189 to 199	As3, Ag1. Ga+, Ld [4] [+]. nig. 2, sicc. 2, strf. 0, elas. 0, struct. homogeneous. lim. sup. 0. Silty clay with a sand and gyttja fraction. Stress structures, no laminae.

Stratum	Height O.D. metres	Depth	Description
8	+0.44 to +0.38	199 to 205	As 4, Ga+, Gs+, Gg. (min.)+, Ld^4 +. nig. 2, sicc. 2, strf. 1, elas. 0, struct. adhesive. lim. sup. 0. Grey silt with a sand and grit fraction, greasy and adhesive. Small organic partings.
7	+0.38 to +0.30	205 to 213	As 4, Lf+, Ld^4 +. nig. 2+, sicc. 2, strf. 1, elas. 0, struct. adhesive. lim.sup. 1. Silt with iron staining and discrete gyttja partings. Grey.
6	+0.30 to +0.21	213 to 222	Ld^4 4, As+, Th (Phra.) ^{2} +. nig. 3+, sicc. 2, strf. 2, elas. +, struct. greasy and adhesive. lim. sup. 0. Gyttja with a slight silt fraction. No inorganic laminae and very occasional monocots. such as *Phragmites.*
5	+0.21 to +0.10	222 to 233	Ld^3 3, Th^2 1, As+, Ag+. nig. 3, sicc. 2, strf. 2, elas. 2, struct. greasy and laminated, lim. sup. 0. Gyttja with clay and silt in discrete laminae and monocots. throughout.
4	+0.10 to +0.02	233 to 241	Ld^4 2, Ag1, As1, Ga+, Lf+, Th^3 +. nig. 2^+, sicc. 2, strf. 1, elas. 0, struct. greasy, lim. sup. 0. Silty clay gyttja with monocot. root partings, iron staining. Slight sand fraction.
3	+0.02 to -0.05	241 to 248	Ag2, Ga1, Ld^4 1, Lf+, Th^3 +. nig. 2+, sicc. 2, strf. 1+, elas. 0, struct. slightly laminated, lim. sup. 0. Silt with gyttja and sand laminae towards top of stratum. Monocot. rootlets, channels iron stained.
2	-0.05 to -0.23	248 to 266	Ag2, As1. Ga1, Lf+, nig. 2, sicc. 2, strf. 2+, elas. 0, struct. laminated, lim.sup. 0. Clayey silt with sandy partings and very occasional iron staining. Laminae of silt and clay.
1	-0.23 to -0.28	266 to 271	Ag.3, Ga1, Lf+. nig. 2, sicc.2, strf. +, elas. 0, struct. slightly laminated, lim. sup. 0. Buff silt with sand laminations and iron staining.

Samples were taken at regular intervals from the biogenic strata for pollen analysis, and the resulting diagram is given as Fig. 9. From this diagram, the following five Local Pollen Assemblage Zones are recognised:

LPAZ	Depth cm.	Zone Characteristics
DM-15a	225	The AP component increases its proportion of

	to 215	the TP through the zone, but never rises above 45% TP. *Quercus* is the most important taxon in this group. The frequency Group D falls throughout the zone, but never below 40% TP: at 225 cm. it accounts for 60% TP. Of the Herbs, grasses dominate the assemblage.
DM-15b	140 to 130	The proportion of the AP Group never exceeds 25% TP: *Quercus* maintains its pre-eminence. Of Group D taxa Grasses maintain values in excess of 65%Σ(AP+C). The proportion of Group F rises from a single grain of 140 cm. to 34% TP at 135 cm., all of which is explained by the behaviour of *Typha angustifolia* which rises to a peak value of 54%Σ(AP+E) at 135 cm.
DM-15c	130 to 117	The proportion of the AP Group rises through the zone, and the AP continues to be dominated by *Quercus*; there are low, persistent values of *Ulmus*. Gramineae values decline through the zone. There is a modest increase in the frequency of the Chenopodiaceae. Of Group E taxa, after the recession of *Typha angustifolia* at the top of zone b, it again rises to 38%Σ (AP+E) at 125 cm. from which level it declines again: it is supported by *Hydrocotyle* and by *Cladium mariscus* which occur as low sporadic values.
DM-15d	60 to 25	There are high *Quercus* values sustained throughout the zone, and a maximum value of 68%ΣAP is recorded at 45 cm. *Ulmus* values tend to fall at the middle of the zone, and then rise steeply towards the top of the zone; at 60 cm. *Ulmus* accounts for 6%ΣAP, at 45 cm. for 2% and at 25 cm. for 21%. The number and frequency of Group D taxa increases: grass values are high at the base of the zone attaining an isolated peak of 77%Σ(AP+D) at 55 cm., from which level they deline and rise slightly at the top of the zone. The Cyperaceae behave in complement attaining a maximum value of 38% at 40 cm. This value would have been higher if *Cladium marisucs* had not been separated. *C. mariscus* rises early in the zone, and values in excess of 50%Σ(AP+E) are sustained from 40 to 45 cm., and thereafter the taxon suffers a decline. A single grain of *Plantago lanceolata* is recorded at 40 cm. and there are sporadic occurrences of *Artemisia, Filipendula* and Chenopodiaceae. At 35 cm. there is a sudden increase in the frequency of fern spores, from 7%Σ(AP+F) at 40 cm. to 64% at 35 cm.
Dm-15e	25 to 5	The proportion of Group F increases while that of Group D is maintained, both at the expense of the AP Group whose proportion falls from 44% TP at 30 to 19% at 24 cm., and thereafter never exceeds 12% of the AP group. *Quercus* values decline and are replaced by both *Alnus* and *Betula*. Coryloid values also rise. *Ulmus*

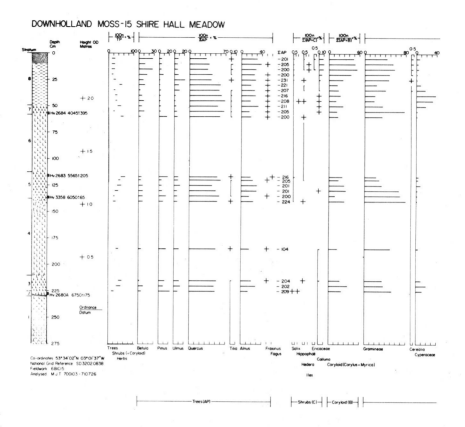

FIG. 9. Pollen diagram from Downholland Moss-15. The percentage frequency of each taxon is based on the calculating formula given for each of the seven groups recognised: AP, Tree taxa: B, Shrub taxa: C, Coryloid: D, Herb taxa: E, Aquatic taxa: F, Spores: G, Varia, comprising unidentified spores, *Pediastrum* colonies and Hystrichosphaeroidae.

Zone Characteristics

values fall from 20% AP at 25 cm. to 8% at 20 cm., and thereafter never exceed 6%: the d/e boundary is drawn at the point where *Ulmus* values fall from 20 to 8%. Grass values sustain high levels, and a small but continuous proportion is contributed by Cerealia. In addition, there is an increase in the frequencies and occurrence of open habitat taxa such as Tubuliflorae, including *Artemisia* and *Centaurea cyanus*, Liguliflorae, Umbelliferae, Rosaceae, including *Filipendula*, *Galium*-type, *Rumex* and *Thalictrum*. The high frequencies of Spores recorded in zone d are sustained throughout zone e.

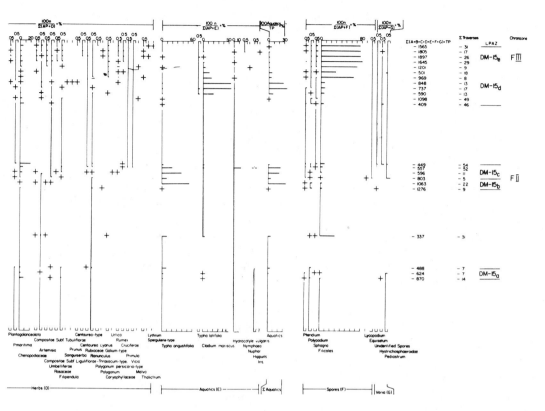

Seven samples were cut from the core and the results of the radiocarbon assays are given below and in Table 1.

No.	Laboratory Code	Years BP	Depth cm	Height O.D. Metres	Stratum
13	Hv. 2680A	6750±175	226 to 229	+0.15	5
14	Hv. 2681	4325±345	213 to 216	+0.30	6
27	Hv. 3357	4695±110	217 to 220	+0.24	6
15	Hv. 2682	4600±430	136 to 140	+1.06	14
28	Hv. 3358	6050± 65	135 to 138	+1.06	15
16	Hv. 2683	5565±205	116 to 119	+1.27	16/15
17	Hv. 2684	4045±395	57 to 60	+1.86	19

The relationship of the dates to the stratigraphic succession is shown on Fig. 7. Dates 14 and 15 were inexplicably young, and two additional assays were obtained. Date 27 was still in error, whilst Date 28 conformed to the series of ages established for the transgression sequences elsewhere. It is probable that the sediments of Stratum 4 were affected by the ground-water regime in the drainage gutter nearby, for the floor of the gutter is close in altitude to zero, Ordnance Datum, and the movement of water has contaminated the sediments in Stratum 4 and the upper and lower parts of Strata 3 and 5 respectively.

Date No. 13, 6750±175, dates the regressive contact of the first transgression recorded on Downholland Moss: a supporting date of 6760±95 at DM-11A derives from material accumulating during the culminating stages of the same transgression further eastward on the moss. For the same time/stratigraphic boundary, the dates are corroborative. The LPAZ at the two sites, DM-11c and DM-15a, are broadly similar in the behaviour and proportions of the taxa represented: in both diagrams, *Pinus* values are highest at the base of the zones and decline thereafter; *Ulmus* values rise and then fall, and this pattern is reflected also in the *Alnus* and Coryloid frequencies. *Quercus* is the dominant taxon in the AP group. There are low frequencies in both LPAZ of Chenopodiaceae and other taxa associated with salt marsh communities.

Both dates 14 and 15 are rejected on the grounds that the sediment from which the dates derived certainly did not accumulate in a post-*Ulmus* stage, and on the grounds of lack of consistency.

The succeeding two dates, Nos. 28 and 16, are accepted again on the grounds of internal consistency, although 28 may be too young because of the presence of *Phragmites* rhizomes. There is a corroborating date from DM-16, east of this sampling site, but there are no independent means of testing unequivocally the accuracy of the dates: certainly the pollen assemblage indicates Stratum 5 was accumulating during Flandrian II.

Date No. 17, 4045±395 is probably too young: the absolute date places the biogenic sediment from Stratum 19 well within Flandrian III, but the pollen assemblage does not corroborate this. The boundary between LPAZ DM-15d and DM-15e is drawn at 25cm where *Ulmus* attains a maximum value of 21 per cent AP, and falls thereafter to 8 per cent. These are still high values, but taken with the continuous record of Cereal pollen and the presence of ruderals, such as *Plantago lanceolata*, *Artemisia*, *Rumex* and *Galium*-type, there can be little doubt that the FII/FIII boundary would be drawn at 25 cm, and a date for this chronozone boundary is given by Hibbert *et al.* (1971) at 5010±80 BP from Red Moss. Date No. 17 is probably in error by more than 1000 radiocarbon years.

Two samples were taken from Strata 16 and 17 for diatom analysis, and the taxa identified from each level are listed below:

105 cm

Nitzschia bilobata Wm. Smith
N. navicularis (de Brebisson ex Kützing) Grunow

N. marina Ralfs in Pritchard
Navicula rhombica Gregory
Diploneis didyma (Ehrenberg) Cleve.
Raphoneis surirella (Ehrenberg) Grunow ex Van Heurck
Coscinodiscus lineatus Ehrenberg
Cyclotella striata (Kützing) Grunow
Paralia sulcata (Ehrenberg) Cleve. (In Werff, *Melosira sulcata* Kützing)
Podosira stellinger (Bailey) Mann.
Actinoptychus senarius Ehrenberg

117 cm

Biddulphia alternans (Bailey) Van Heurck
Diploneis didyma (Ehrenberg) Cleve.
Coscinodiscus lineatus Ehrenberg
Paralia sulcata (Ehrenberg) Cleve.
Actinoptychus senarius Ehrenberg
Podisira stelliger (Bailey) Mann.
Nitzschia navicularis (de Brebisson ex Kützing) Grunow
N. punctata Wm. Smith

At these two levels, on the basis of the ecological preferences of the taxa, the environmental conditions of the sedimentary environment can be assessed approximately. Van der Werff and Huls (1958–66) have established criteria for calculating the marine-brackish–freshwater ratio on the basis of the included diatoms. At 105 cm, the marine element accounts for 44 per cent while the marine-brackish element accounts for 22 per cent and the brackish element for 33 per cent. At 117 cm the marine element has risen to 62 per cent and 12 per cent each for the marine-brackish, brackish marine and brackish element (see Appendix, pp. 203–8).

Ten bulk samples were taken from Strata 19-16, 14-7 and 4-1 and the particle size distribution calculated for each level following standard methods of coarse analysis by dry sieving and fine analysis by pipette Method. The results are given in Table 2, and shown graphically in Tooley (in the press).

Table 2. Downholland Moss-15. Particle Size Distribution

Particle Size Type	Class Limits (mm)	60cm	75	85	105	150	170	185	205	240	265
Fine Gravel	6.0 -2.0					0.09					
Coarse sand	2.0 -0.6	0.25%	0.09	0.06	0.05	0.23	0.10	0.05	0.05		
Medium sand	0.6 -0.2	0.13	0.14	0.11	0.05	0.05	0.10	0.43	0.09		0.05
Fine sand	0.2 -0.06	0.19	0.23	0.11	0.10	0.05	0.41	0.96	0.28	0.11	0.23
Coarse silt	0.06 -0.02	8.42	6.86	3.95	3.16	25.33	48.17	19.62	7.28	26.03	51.29
Medium silt	0.02 -0.006	24.94	28.25	28.02	18.00	26.55	21.62	22.85	23.65	21.18	20.12
Fine silt	0.006-0.002	18.39	60.22	15.44	18.49	13.63	7.21	23.74	23.19	17.92	7.95
Clay	<0.002	47.69	4.22	52.32	60.15	34.07	22.39	39.34	45.46	34.76	20.36

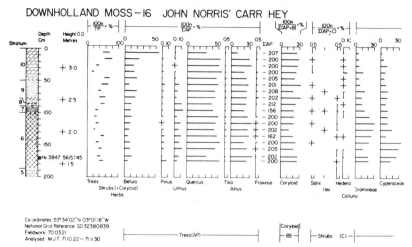

FIG. 10. A pollen diagram from Downholland Moss-16. The percentage frequency of each taxon is based on the calculating formula for each of the eight groups recognised, which comprise: AP, Tree taxa: B, Coryloid: C, Shrub taxa: D, Herb taxa: E, Aquatic taxa: F, Pteridophyte and Bryophyte taxa: G, Fungal spores: H, Algae.

The undried weight of the samples used ranged from 21.04 to 24.53 g. The sediments from these Strata were dominated by the silt and clay fractions, the proportions of which varied through the strata analysed.

Downholland Moss—16, DM-16, John Norris' Carr Hey.

A fourth site on Downholland Moss was chosen at DM-16 where an upstanding baulk of peat allowed sampling from the deepest remaining surface peat on the Moss. It was hoped that this site would provide evidence for post-elm decline episodes on the Moss. Samples were taken with a Russian-type sampler for pollen and radiometric analysis. The stratigraphy was recorded in the field, from samples taken with a Hiller in 1968:

Stratum	Height O.D. metres	Depth	Description
10	+3.20 to +2.70	000 to 050	Str.conf.Sh [4] 4 nig. 4, strf. 0, elas, 1, sicc. 3. Peaty soil, giving way to a black, amorphous peat.
9	+2.70 to +2.35	050 to 085	Sh [4] 4, Dl.+ nig. 4, strf. 0, elas. 1, sicc. 3, lim.sup.0. Black, amorphous peat with some *Alnus* wood at 75 to 80 cm.
8	+2.35 to +2.30	085 to 090	Th [1] *(vagi.)* 4, Th [0] *(Meny.)*+ Dh+. nig. 3, strf. 0, elas. 1, sicc. 3, lim.sup.0. Partly humified *Eriophorum* peat, with *Menyanthes*

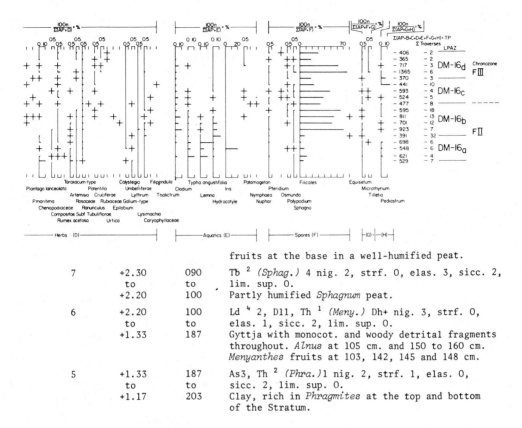

fruits at the base in a well-humified peat.

7	+2.30 to +2.20	090 to 100	Tb ^{2} *(Sphag.)* 4 nig. 2, strf. 0, elas. 3, sicc. 2, lim. sup. 0. Partly humified *Sphagnum* peat.
6	+2.20 to +1.33	100 to 187	Ld ^{4} 2, D11, Th ^{1} *(Meny.)* Dh+ nig. 3, strf. 0, elas. 1, sicc. 2, lim. sup. 0. Gyttja with monocot. and woody detrital fragments throughout. *Alnus* at 105 cm. and 150 to 160 cm. *Menyanthes* fruits at 103, 142, 145 and 148 cm.
5	+1.33 to +1.17	187 to 203	As3, Th ^{2} *(Phra.)*1 nig. 2, strf. 1, elas. 0, sicc. 2, lim. sup. 0. Clay, rich in *Phragmites* at the top and bottom of the Stratum.

Samples were taken from the cores for pollen analysis from Strata 6, 7, 8, 9 and 10 at 10 cm intervals. The pollen diagram is shown as Fig. 10, and four Local Pollen Assemblage Zones are recognised:

LPAZ	*Depth* *cm.*	*Zone Description*
DM-16a	174 to 135	The AP proportion of the TP never exceeds 45%, and the assemblage is characterised by a preponderance of Group D taxa of which grasses and sedges are the most conspicuous. Chenopodiaceae values decline sharply, and there are low, sporadic frequencies of Compositae, particularly *Artemisia*. The AP group is dominated by *Quercus* and *Alnus*. *Ulmus* attains a maximum value of 13%ΣAP at 155cm. The proportion of Group E taxa rises suddenly at the base of the zone, attains a maximum value of 17% TP at 155 cm. and declines slowly thereafter. Taxa contributing to this frequency distribution

include *Typha angustifolia, Lemna* and *Iris.*

DM-16b	135 to 85	Taxa from Groups D and F dominate the assemblage. Grass values decline from moderate frequencies at the base of the zone, and there is a modest decline in sedge values. Fern spores rise suddenly at the base of the zone and maintain high values throughout the zone, declining only at the top of the zone at 135 cm. Filicales account for 8% Σ(AP+F) and at 125 cm. this has risen to 71%. There are low values of *Cladium mariscus* and *Hydrocotyle* declines after a maximum value of 13%Σ(AP+E) at the base of the zone. The AP group continues to be dominated by *Quercus* and *Alnus. Pinus* becomes discontinuous, and *Ulmus* falls during the first part of the zone, recovers to attain peak values of 14% at 95 cm., and falls suddenly at 85cm. The b/c boundary is drawn at the point where *Ulmus* values fall from 12%ΣAP at 85 cm. to 3% at 75cm.
DM-16c	85 to 45	The AP group taxa increase their proportion throughout the zone, and continue to be dominated by *Quercus* and *Alnus.* Both *Pinus* and *Ulmus* record low but continuous frequencies. *Corylus* values rise at the base of the zone and fluctuate thereafter. There are low, sporadic frequencies or individual records of *Plantago lanceolata, Rumex, Ranunculus, Filipendula* and *Pteridium.* Filicales rise to a second maxima of 49%Σ(AP+F) at 65 cm. from which level they then decline to the top of the zone.
DM-16d	45 to 5	The zone is characterised by a collapse of AP group values at the base of the zone and their subsequent recovery. *Quercus* and *Alnus* continue to be the most conspicuous taxa, but both decline as *Betula* frequencies rise at the top of the zone. The behaviour of the AP group complements that of Group F taxa, and, in particular, the Filicales which rise steeply from 18%Σ(AP+P) at 45 cm. to 77% at 35 cm. and decline thereafter.

A 5 cm thick sample for radiocarbon dating was taken from the lower part of Stratum 6 and gave a date of 5615±45 BP (Table 1, No. 42), which places this stratum well within Chronozone FII. The FII/FIII boundary is drawn at 80 cm, where the *Ulmus*-decline is under way: the boundary is tentative, for there is no absolute date at this level. The relationships between the LPAZs at DM-16 and the RPAZs at Red Moss is shown on Fig. 4.

Hillhouse

The area west of Hillhouse Farm (Figs. 2 and 11) was selected for detailed stratigraphic and pollen analysis (Tooley 1976b), because it had been chosen as the type locality for the *Hillhouse coastline* by Gresswell (1953) (see Fig. 14).

Sampling sites were selected at points where a break of slope was evident in the field, and, although this produced an irregular sampling interval, it was consistent with Gresswell's field practice of mapping slope facets and slope inflexions.

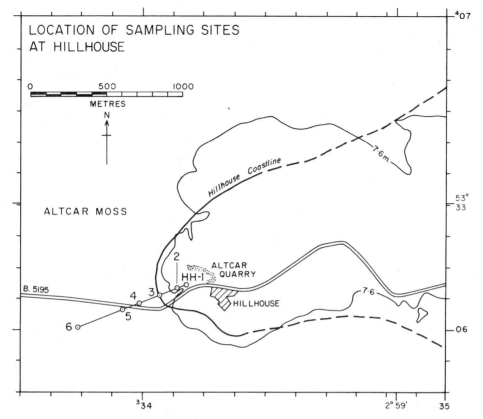

FIG. 11. A map to show the location of sampling sites at Hillhouse, across the 'Hillhouse Coast-line'. The line of the coast has been taken from Gresswell (1953): a continuous line represents 'a slope but no definite feature' whereas a pecked line indicates 'no feature'. Sampling sites are indicated by a circumscribed point, and were chosen where there was a break of slope or slope inflexion.

The stratigraphic succession was recorded at the six sampling sites and the results are shown on Fig. 12. The deepest biogenic succession was recorded at HH-6, where the following eight strata were recorded:

Stratum	Height O.D. metres	Depth	Description
8	+2.55 to +2.05	000 to 050	Str. conf. Black organic soil.
7	+2.05 to +1.72	050 to 083	Th ^{3} 4 Well-humified monocot. peat.

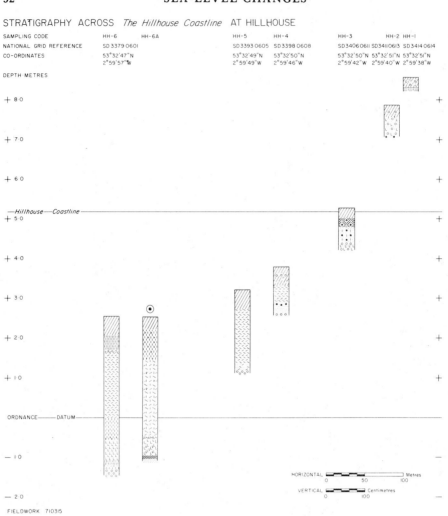

STRATIGRAPHY ACROSS *The Hillhouse Coastline* AT HILLHOUSE

FIG. 12. The stratigraphy across the 'Hillhouse Coastline' at Hillhouse. The mean height of the 'Hillhouse Coastline' given by Gresswell (1953, 1957) at +5.15 metres O.D. is shown by a horizontal line.

| 6 | +1.72 to -0.10 | 083 to 265 | D1.4, Ld [4] + Woody, detrital peat, with a mud fraction towards the base. Charcoal at 90 cm. |
| 5 | -0.10 to -0.57 | 265 to 312 | Th [3] 4, Th [1] (*Phra.*)+. D1.+ Well-humified monocot. peat, with some woody detrital fragments, becoming fresher with *Phragmites* towards the base. |

4	-0.57 to -1.12	312 to 367	As2, Ag2, Sh [4] + Tenacious, blue-grey, silty clay with occasional organic partings throughout, and rich in fresh organic material towards Strata 3 and 5.
3	-1.12 to -1.21	367 to 376	Th [3] 4 Well-humified, compact, monocot. peat.
2	-1.21 to -1.27	376 to 382	Ga3, Sh [4] 1 Organic black sand.
1	-1.27 to -1.45	382 to 400	Gs4 Coarse, grey sand, becoming impenetrable.

Samples for pollen analysis were taken by Dr. P. Cundill in February 1974
from a site adjacent to HH-6, where the following stratigraphic succession was
recorded:

Stratum	Height O.D. metres	Depth	Description
5	+1.44 to -0.46	111 to 301	Mid-brown monocot. peat with wood remains. *Betula* and a few *Phragmites* rhizomes above 161 cm.
4	-0.46 to -0.52	301 to 307	Transition.
3	-0.52 to -0.94	307 to 349	Blue grey clay with a few organic remains. *Phragmites*, especially towards base.
2	-0.94 to -1.02	349 to 357	Gyttja with blue-grey clay and macro-fossils, wood and *Phragmites* rhizomes.
1	-1.02 to -1.06	357 to 361	Mid-grey, coarse sandy material, becoming impenetrable towards the base.

Nine samples were available for pollen analysis from Strata 2, 4 and 5, and
the pollen diagram is shown in Fig. 13. Three Local Pollen Assemblage Zones
are recognised:

LPAZ	Depth cm.	Zone Characteristics
HH-6_a	358 to 352	*Alnus, Pinus* and *Quercus* are the characteristic tree taxa. Group A accounts for over 40%ΣG throughout the zone of the herb taxa, Chenopodiaceae rise throughout the zone.

HILLHOUSE −6A

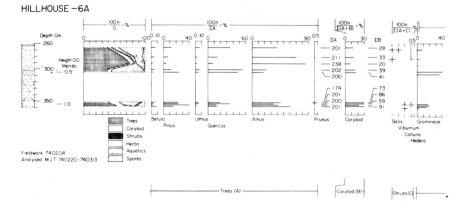

Fig. 13. A pollen diagram from Hillhouse-6A. Six taxonomic groups are shown: A, Tree taxa: B, Coryloid: C, Shrub: D, Herb: E, Aquatic: F, Bryophyte and Pteridophyte taxa. The percentage frequency at successive levels is derived from the calculating formulae given for each group.

| HH-6$_b$ | 352 to 301 | *Quercus, Pinus* and *Alnus*. Group A taxa account for less than 40%ΣG, whereas Group D taxa contribute over 50%ΣG, of which Gramineae and Cyperaceae are most conspicuous. A boundary is drawn at 301 cm. where A values attain that highest frequency and *Alnus* records a peak value. |
| HH-6$_c$ | 301 to 266 | The characteristic tree taxon is *Alnus*. Group A taxa rise to a maximum frequency of 76%ΣG at 291 cm. and decline thereafter Group E taxa increase in frequency throughout the zone, and achieve maximum values at the top of the zone. |

On the basis of the concurrence of the LPAZ with the RPAZ at Red Moss, Strata 2, 3 and 4 may be assigned to the transition of Flandrian I/II and the early part of Flandrian II, whereas Stratum 5 is almost certainly entirely within Flandrian II.

The complexity of the sedimentary environments in the Hillhouse area and adjacent sites had been demonstrated by de Rance (1868) Reade (1871) and Gresswell (1957), based on the stratigraphic descriptions of Hall (1954-5). Fig. 14 shows the distribution of post-glacial deposits in this area according to these authors. The implications of these data for the age and origin of the *Hillhouse coastline* are considered in Chapter 4 and Tooley (1976b).

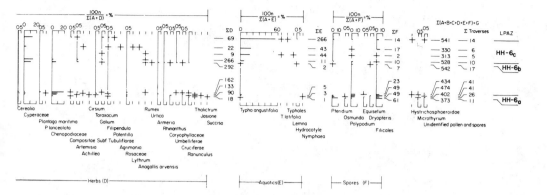

LANCASHIRE: FYLDE IN AMOUNDERNESS

The Keuper Marls and Sandstones that underlie the whole of Amounderness do not outcrop anywhere, and their existence and extent have been proved by borings. The solid is buried by a mantle of till, the thickness of which ranges from 3 to 77 metres. The base of the till occurs from −1.5 metres to −67.1 metres O.D. The till surface also undulates: it occurs as a modest ridge +7.0 to +18.0 metres O.D., running from north to south and reaching maximum elevations in Piper's Heights, Peel, Lower Ballam and Lytham Hall Park (Fig. 20). The ridge continues south at a reduced altitude and is buried by blown sand until it outcrops as a scar on the north bank of the River Ribble, where it is known as Church Scar and is shown on Fig. 30. The mosslands behind Lytham and St. Annes are interrupted by low, rounded knolls of till about +7.0 metres O.D.: flanking Great Marton Moss, there are till outliers, such as Midgeland, and Wild Lane follows a till salient northwards from Lytham Moss.

Elsewhere, the till surface is obscured by deposits of Late Devensian and Flandrian age. Basins in the till have been filled with clays, silts, gyttjas and peats up to 18 metres thick. Some of the basins in the till, such as Marton Mere, are enclosed and sedimentary units derive in part from slope slumping and in part from autochthonous biogenic accumulation (Tooley, unpublished). Towards the coast, basins beneath Lytham Moss, Lytham Hall Park and Nancy's Bay have been affected by marine transgressions, and biogenic sedimentation has been interrupted on several occasions during the Flandrian stage.

Within Lytham Township in the Fylde of Amounderness five sites have been selected for detailed study: Nancy's Bay, Lytham Hall Park, The Starr Hills, Lytham Common and Lytham Moss.

Similar stratigraphic successions have been proved in Over Wyre (Dr. B. Barnes, unpublished); the results of preliminary work on Cockerham and Pilling Mosses have been published by Oldfield and Statham (1965).

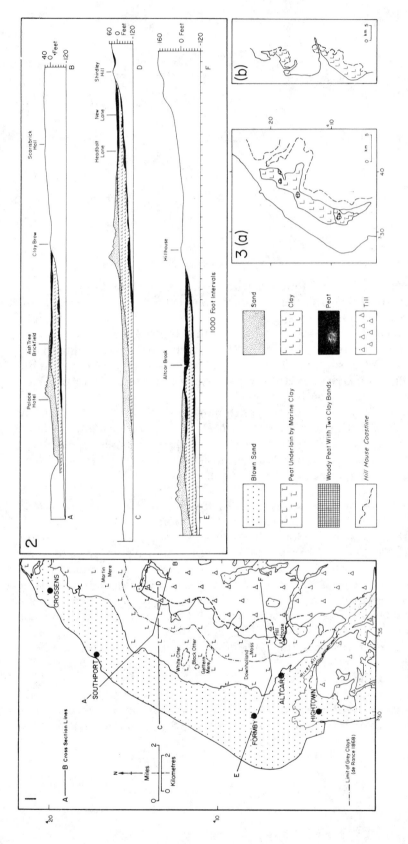

FIG. 14. Postglacial deposits in south-west Lancashire.

14.1. Superficial geology map of south-west Lancashire after de Rance (1868), Reade (1871) and the Soil Survey of England and Wales.

14.2. General sections from south-west Lancashire, redrawn from Reade (1871); the locations are shown on Fig. 14.1.

14.3a. Shows the distribution of lagoonal and tidal flat deposits in south-west Lancashire, based on Hall (1955, 1954–5) and the extent of the Hillhouse Coastline (Gresswell 1957).

14.3b. Shows the extent of the Hillhouse Coastline in south-west Lancashire, the Fylde, Over-Wyre and Lonsdale (Gresswell 1964).

Nancy's Bay, Lytham

Nancy's Bay formerly occupied the low-lying area east of Saltcotes at the East End of Lytham, and west of Warton Hall (Fig. 15). The mouth of the Bay was contracted in the vicinity of Dock Bridge by shingle, but became dilated landward to the north-east towards Carr Farm and north-west towards Eastham Bridge. The area of the Bay is shown on Fearon and Eye's Survey of the River Ribble in 1736. It is named in 1720 and again in 1836 and 1838 (Barron 1938), but there is evidence that it ceased to exist as an embayment covered at each tide in 1742 (see Tooley 1977a).

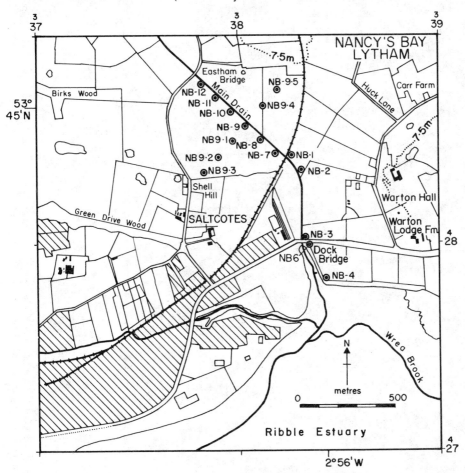

FIG. 15. A map of Nancy's Bay, Lytham, to show the positions of the sampling sites.

The floor of the former Bay lies at altitudes little above +1.5 metres O.D., rising both northwards to the mosses and seawards to Warton Outmarsh, where intertidal sedimentation has carried the altitude of the upper marsh to

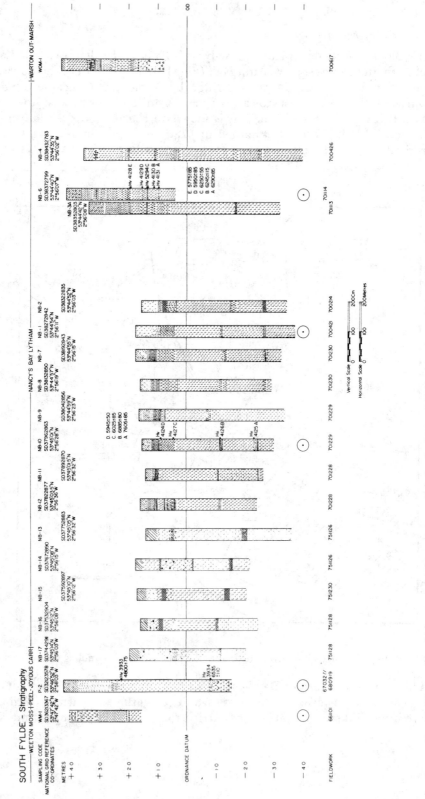

Fig. 16. The stratigraphic successions of fourteen sampling sites in the South Fylde. Sites from which samples were taken for pollen analysis are indicated at the base of the stratigraphic column by an encircled dot.

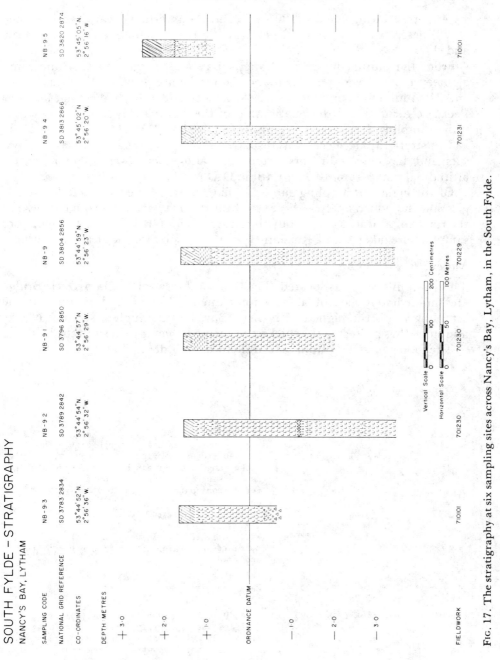

Fig. 17. The stratigraphy at six sampling sites across Nancy's Bay, Lytham, in the South Fylde.

+4.4 metres during the last 100 years and the intertidal peat that de Rance (1872) described is now buried beneath 100 cm of pebbles and estuarine silt and clay.

From the outmarsh, across Nancy's Bay and normal to this line some eighteen borings were put down and the results are displayed in Figs. 16, 17 and 21. Additional borehole records have been added from Weeton Moss and Peel-2 to demonstrate the relationship of the Nancy's Bay series to representative sampling sites in the Lytham-Skippool valley (Fig. 21).

The stratigraphic successions displayed show an alternating series of clays, silts and biogenic sediments extending, altogether, over an eight metre altitudinal range from –4.0 to + 4.0 m O.D.

Of the eighteen sampling sites, two have been selected for detailed stratigraphic and pollen analysis, because of the diversity of stratigraphic types and the range of altitudes covered by the strata. From NB-10 and NB-6 half-core samples were taken with a Russian-type peat sampler for radiocarbon dating.

Nancy's Bay—6. NB-6

This sampling site was located (Fig. 15) south-east of Lytham Dock Bridge and immediately east of a sea embankment. The ground altitude at the sampling site is the highest in Nancy's Bay, and the highest stratum of clay and stones has protected the thickest surviving biogenic deposit in the bay.

The following stratigraphic succession was recorded:

Stratum	Height O.D. metres	Depth	Description
13	+4.15 to +3.65	000 to 050	Str. conf. Surface Soil. Pasture. A tough clay with rounded stones, difficult to penetrate by bucket auger. Underlain by coarse sand at 20 cm. passing into tenacious, buff clay with ferruginous partings. Sharp stratigraphic boundary at 50 cm.
12	+3.65 to +3.35	050 to 080	D1.4, nig. 4, strf. 0, elas. 0, sicc. 1, lim. sup. 4. Well-humified, black, woody detrital peat: oxidised and crumbly above, fragments of bark and wood below.
11	+3.35 to +3.25	080 to 090	Th ^{2} 4 Partly humified monocot. peat.
10	+3.25 to +2.45	090 to 170	D1.4, Dh+, Th ^{2} (*Phra.*)+, Gg (maj.)+. Partly humified, woody, detrital peat: *Betula* at 118 to 125 cm. and at 150 cm. Monocot. stems at 121 cm. and *Phragmites* rhizomes at 145 cm.
9	+2.45 to +1.92	170 to 223	Th ^{2} 3, Th ^{1} (*Phra.*)1. Partly humified, monocot. peat, with *Phragmites* at 170 to 172, 180 to 189, 200 and a *Phragmites* peat at 220 to 223.

Stratum	Height O.D. metres	Depth	Description
8	+1.92 to +1.54	223 to 261	As3, Sh⁴ 1, Th¹ (Phra.)+. Blue clay, with black organic partings. Transition zone above, 4 cm.: buff to blue clay with unhumified *Phragmites* rhizomes.
7	+1.54 to +1.39	261 to 276	Th² (Phra.)1. Partly humified, *Phragmites* peat.
6	+1.39 to +1.33	276 to 282	Ld³ 3, Th² (Phra.)1. Mud, rich in *Phragmites*.
5	+1.33 to +1.15	282 to 300	As2, Sh⁴ 1, Th² (Phra.)1. Blue clay, with black, organic partings, becoming buff and rich in *Phragmites* at the base.
4	+1.15 to +1.00	300 to 315	Th² (Phra.)4. *Phragmites* peat, partly humified, especially towards the base. Dry and fissile.
3	+1.00 to +0.95	315 to 320	Ld³ 2, Th² (Phra.)2 D1+ Mud, rich in *Phragmites* and with woody, detrital fragments.
2	+0.95 to +0.87	320 to 328	As2, Ag2, Th¹ (Phra.)+. Coarse, silty, grey-blue clay with a dense unhumified, *Phragmites* band at 327 cm.
1	+0.87 to +0.65	328 to 350	Ag2, Ga2. Coarse, sandy grey, blue silt.

Samples for pollen analysis were taken from the core at regular intervals from Strata 3, 4, 6, 7, 9, 10, 11 and 12. A pollen diagram has been constructed and is displayed in Fig. 18. The following six Local Pollen Assemblage Zones have been identified:

LPAZ	Depth cm.	Zone Characteristics
NB-6a	313 to 303	The summary diagram indicates that Herb taxa dominate the assemblage accounting for some 60% TP. Tree taxa increase their proportion marginally through the zone. Of the tree taxa *Quercus* and *Pinus* are the most important, the former rising, the latter declining in frequency throughout the zone. Grass pollen dominates the herb taxa, and accounts for over 30%Σ(AP+D). There are low frequencies of Compositae pollen, and low values of Chenopodiaceae. Aquatic taxa account for up to 3% TP, and the highest proportion of this is contributed by *Typha angustifolia*.
NB-6b	279 to 265	The AP proportion of the TP declines, whilst the herb proportion increases through the zone. *Quercus* values are high, reaching a maximum

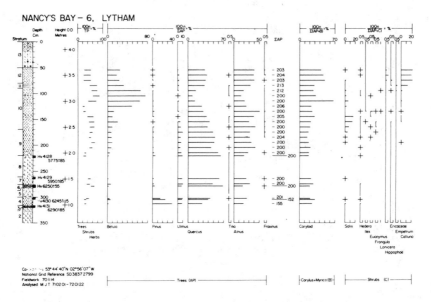

FIG. 18. A pollen diagram from Nancy's Bay—6. The frequency of each taxon at successive levels through the biogenic deposit is based on the calculating formula given for each group. Eight groups are recognised: AP, Tree taxa: B, Coryloid group: C, Shrub taxa: D, Herb taxa: E, Aquatic taxa: F, Pteridophyte and Bryophyte taxa: G, Algae: and H, Varia.

frequency of 69%ΣAP at 279 cm. *Alnus* values are moderate. *Pinus* values are very low compared to zone a. Of the herb taxa, the Chenopodiaceae remain with persistent but low values. The proportion of aquatic taxa rises from 10%Σ(AP+E) at 279 cm. to a maximum of 13% at 275 cm.: a peak value of 10%Σ(AP+E) for *Typha angustifolia* coincides with high values of *Lemna*, for example 18%Σ(AP+E) also at 275 cm.

NB-6c	221
	to
	175

Quercus and *Alnus* account for over 75% AP throughout the zone. *Pinus* is sporadic. *Ulmus* values fall at first and then rise: at 221 cm. they are 9%ΣAP, and fall to 4% at 205 cm., but rise again to 10% at 175 cm. Grass values are high at the opening of the zone, fall considerably and rise towards the top of the zone. Chenopodiaceae values fall from a peak of 9%Σ (AP+D) at the opening of the zone. The proportion of aquatic taxa is high at the opening of the zone, 19% TP at 215 cm. falling thereafter but rising again to 12% TP at 185 cm. The first peak is accounted for by high values of *Lemna* which accounts for 39%Σ(AP+E) at 215 cm., and by *Cladium* at 185 cm., with 20%Σ(AP+E).

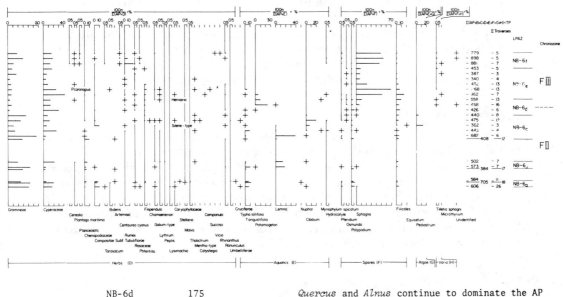

| NB-6d | 175
to
145 | *Quercus* and *Alnus* continue to dominate the AP values. *Quercus* begins to decline slightly, and this is sustained through to zone e. Both *Ulmus* and *Pinus* become discontinuous, and this zone contains the secondary *Ulmus* decline: at the opening of the zone elm values are 10%ΣAP, and they fall to 5% at 155 cm., while a single grain is recorded at 145 cm. *Salix* values rise to a maximum of 16%Σ(AP+C) at the top of the zone. Of the aquatic taxa, *Potamogeton* rises suddenly to 24%Σ(AP+E) at 155 cm. from which levels it then declines. There is the beginning of a continuous Sphagna frequency. |

| NB-6e | 145
to
85 | This zone is characterised by high frequencies of *Betula*, Coryloid and Sphagna, all of which rise, sustain high values and then decline. Both *Quercus* and *Alnus* values are depressed. *Betula* accounts for 75%ΣAP at 105 cm. which peak is anticipated at 115 cm. by a Coryloid peak value of 40%Σ(AP+B). Sphagna peaks at 125 cm. with 69%Σ(A+F). |

| NB-6f | 85
to
56 | *Betula* values decline whilst those of *Alnus* and *Quercus* rise, although they never regain the levels achieved in earlier zones. Coryloid frequencies rise and maintain moderate values. There is a rise in the frequency of dwarf shrubs, especially *Calluna*, together with low frequencies of *Hippophae*. The first ruderals appear, such as *Plantago lanceolata*, *Taraxacum*-type, *Centaurea cyanus*, and *Rumex*. *Rumex* achieves a peak value of 18%Σ(AP+D) at 75 cm. This peak coincides with a low frequency of Cerealia. Sphagna, after low values across the e/f zone boundary, rise strongly and sustain values greater than 45%Σ (AP+F) for the rest of the zone. |

Two chronozones are recognised, although the boundary, drawn at 160 cm, must be tentative, because of the absence of an absolute date here. The biogenic deposits that have been dated radiometrically place Strata 3 and 9 unequivocally within the latter part of chronozone Flandrian II. The boundary is drawn at the point where the elm decline is under way, although at this level there is no evidence at all for anthropogenic influence at the site.

Nancy's Bay—10, NB-10

The second site examined is located within the north-western quadrant of Nancy's Bay. The silt and clay facies recorded here were thicker than those at sites to seaward where the ground surface rose, and the biogenic facies, though thin, occurred much deeper than elsewhere. Taken with NB-6, it provides evidence for the age of all the transgressions recorded in Nancy's Bay. A half-core sample was taken with a Russian-type sampler, from which samples for pollen analysis and radiocarbon dating were taken. The following stratigraphic succession was recorded:

Stratum	Height O.D. metres	Depth	Description
11	+1.52 to +1.22	000 to 030	Str. conf. Brown clay, with peaty partings and rounded stones above, passing down into black, oxidised peat.
10	+1.22 to +1.02	030 to 050	Sh ^{4} 4. Oxydised peat.
9	+1.02 to +0.87	050 to 065	Th ^{2} 4. Partly humified, monocot. peat, with occasional blue, beetle elytra.
8	+0.87 to +0.76	065 to 076	As3, Th ^{2} (Phra.)1. Blue clay, rich in *Phragmites*, with a 4 cm. transition zone above, and 3 cm. below.
7	+0.76 to +0.45	076 to 107	Th ^{2} (Phra.)4. Partly humified *Phragmites* peat.
6	+0.45 to +0.48	107 to 200	As3, Sh ^{4} 1, Th ^{2} (Phra.)+. Ag+. Grey-blue clay, becoming buff towards base, with black organic partings and silty partings. A 4 cm. transition zone above, with partly humified *Phragmites* partings.
5	-0.48 to -1.14	200 to 266	Ag4. Sh ^{4} +. Grey silt, with very occasional organic partings, becoming coarse, and running in places. 3 cm. transition below.
4	-1.14 to -1.27	266 to 279	Th ^{3} 4. Th ^{2} (Phra.)+. Well-humified, monocot. peat with some *Phragmites*.

3	-1.27 to -2.40	279 to 392	Ag4. Sh [4] +. Th [2] *(Phra.)+*. Blue-grey silt, coarse in places with black organic partings at 375 cm. to 376 cm., and finer, clay partings. A 7 cm. transition zone with *Phragmites* rhizomes above and an 8 cm. transition below.
2	-2.40 to -2.48	392 to 400	Th [4] 4. Completely humified, monocot. peat. Sharp contact at 392 cm.
1	-2.48 to -2.98	400 to 450	Ag4. Running, wet, coarse grey-blue silt, with no organic partings. Impossible to penetrate further.

Samples for pollen analysis were taken from Strata 2, 4, 7, 9 and 10, but in the lowest biogenic stratum (2) poverty of numbers prevented more than one spectrum from being counted. A pollen diagram from NB-10 is given as Fig. 19. Four Local Pollen Assemblage Zones are recognised:

LPAZ	*Depth* *cm.*	*Zone Characteristics*
NB-10a	392	The zone is characterised by high *Pinus* values which account for 60%ΣAP. *Quercus* values are more than double those of *Ulmus*, while *Alnus* has a frequency of 7%ΣAP. Of the herbs, Gramineae are the dominant taxon. Open habitat taxa such as *Filipendula*, *Taraxacum*-type and *Pteridium* occur in low frequencies.
NB-10b	275 to 265	Tree taxa have a low proportion of the TP, but this proportion increases towards the top of the zone. *Pinus* values are considerably reduced compared to zone a, whilst those of *Quercus* rise to a maximum value of 56%ΣAP at 267 cm. Chenopodiaceae frequencies are continuous and reach a peak value of 16%Σ(AP+D) at 267 cm. Of the aquatic taxa, *Lemna* is dominant rising to 35% Σ(AP+E) at 267 cm.
NB-10c	100 to 65	*Quercus* values are high and decline through the zone, whilst those of *Alnus* rise steeply at the top of the zone. Grass frequencies decline throughout, whilst those of the Chenopodiaceae are continuous at first and then become sporadic. The feature of the zone is the rise in the frequency of *Typha angustifolia*, and its subsequent decline: at 85 cm. it accounts for 62%Σ(AP+E).

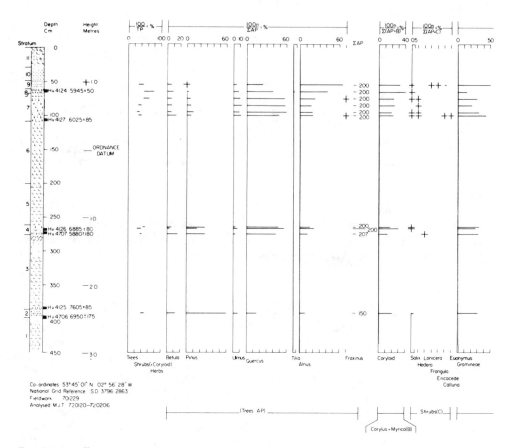

FIG. 19. A pollen diagram from Nancy's Bay — 10. The percentage frequency of each taxon at successive levels is based on the calculating formula given for each group: seven groups are recognised, AP, Tree taxa: B, Coryloid: C, Shrub taxa: D, Herb taxa: E, Aquatic taxa: F, Pteridophyte and Bryophyte taxa: G, Varia, comprising Hystrichosphaeroidae and *Microthyrium*.

NB-10d	65	*Alnus* rises strongly, but all other tree taxa
	to	decline. *Ulmus* values fall from 9%ΣAP at 65 cm.
	55	to 4.5% at 55 cm.: *Pinus* becomes discontinuous.
		There are marked increases in the frequencies
		of Gramineae, *Hydrocotyle* and Sphagna.

The relationship of the Local Pollen Assemblage Zones NB-10 a-d to the Regional Pollen Assemblage Zones at Red Moss is shown in Fig. 4. On the basis of the range of absolute ages given by the radiocarbon dates, two chronozones can be recognised, although the boundary drawn at 350 cm is

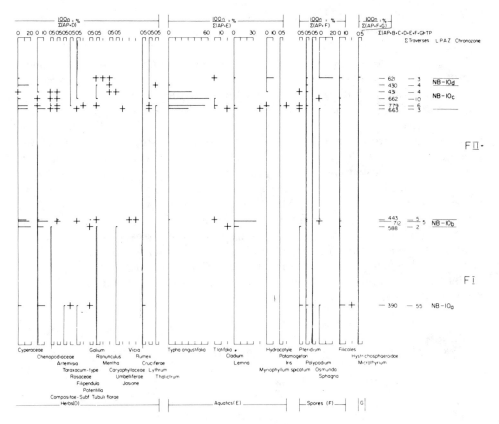

speculative. FI comprehends Strata 1, 2 and 3, and FII Strata 3, 4, 5, 6, 7 and 9.

Lytham-Skippool Valley

Bisecting the Fylde is a narrow valley some 14 km long joining Skippool, on the River Wyre, in the north, and Lytham, on the River Ribble, in the south. At its narrowest the valley is about 0.3 km wide. The floor of the valley undulates, and ground altitudes range from a maximum of +7.07 metres on the upstanding baulk of Weeton Moss (SD 36213367) to a minimum value on the reclaimed floor of Nancy's Bay in the south of +1.6 at Nancy's Bay-12. The valley opens southwards towards Lytham, where low till outliers rise above the reclaimed mosses, and becomes confined in its central and northern section, where it is flanked by a low gently rolling till landscape with kettle holes and other dead-ice features. The drainage pattern is artificial, by way of drains and gutters: the artificial watershed is immediately north of sampling site MyM-3 (Fig. 20).

In order to prove the extent and age of the marine clay that had been demonstrated in Nancy's Bay, a series of exploratory borings was put down at approximately one kilometre intervals between the limits of Nancy's Bay, Lytham and Skippool on the Wyre estuary. The location of the sampling sites is shown on Fig. 20 and the stratigraphy on Fig. 21. Pollen diagrams have been drawn up from Nancy's Bay, Peel and Weeton Moss, and the first two sites have been described in detail here. At only one site within the valley, Peel-2, was it possible to reach the till subsurface at about −2.5 metres O.D.

In 1967 samples for pollen analysis were obtained using a Hiller-type peat sampler, but it was only possible to penetrate silty clay to 450 cm (−0.25 metres O.D.). Subsequently, in 1968 a piston core sample was obtained of the lower peat, and samples from this core were taken for radiocarbon dating and to complement the earlier pollen diagram. The location and stratigraphy of the sampling site at Peel (P-2) are shown on Figs. 20 and 21.

The stratigraphy is given below; Strata 1-5 were recorded in 1968 and Strata 6-9 in 1967.

Stratum	Height O.D. metres	Depth cm.	Description
9	+4.25 to +3.75	000 to 050	Str. conf. Top soil. Black oxidised peat. Arable.
8	+3.75 to +2.63	050 to 162	D14 Crumbly, brown peat, becoming fresher towards the base, and with woody remains throughout.
7	+2.63 to +2.34	162 to 191	Th^2 *(Phra.)* 4 *Phragmites* peat.
6	+2.34 to +2.25	191 to 200	As4 Th^0 *(Phra.)*+ Smooth, tenacious, blue-grey clay with a 3cm. transition above, with fresh *Phragmites* rhizomes.
5	+2.25 to -0.25	200 to 450	As3 Ag1 Blue-grey clay, becoming progressively more silty.
4	-0.25 to -0.80	450 to 505	Ga4 As+ Th^2 *(Phra.)*+ Fine, grey sand with a clay fraction and some *Phragmites* above.
3	-0.80 to -1.07	505 to 532	D13 Th^3 1 Th^1 *(Phra.)*+ Well-humified, woody detrital peat with an increasing monocot. fraction

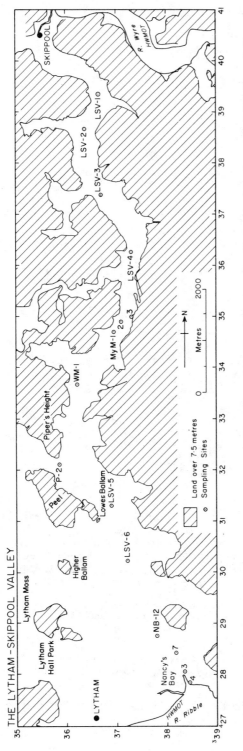
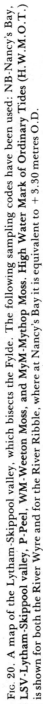

FIG. 20. A map of the Lytham-Skippool valley, which bisects the Fylde. The following sampling codes have been used: NB-Nancy's Bay, LSV-Lytham-Skippool valley, P-Peel, WM-Weeton Moss, and MyM-Mythop Moss. **High Water Mark of Ordinary Tides (H.W.M.O.T.)** is shown for both the River Wyre and for the River Ribble, where at Nancy's Bay it is equivalent to +3.30 metres O.D.

THE LYTHAM-SKIPPOOL VALLEY: Stratigraphy

SAMPLING CODE	NB-4 NB-3A	NB-7	NB-12	LSV-6	LSV-5	P-2,P-2A
NATIONAL GRID	SD3843 2783	SD3819 2843		SD 3721 3022	SD 3689 3129	SD 3212 3578
REFERENCE	SD 3835 2803		SD 3782 2877			

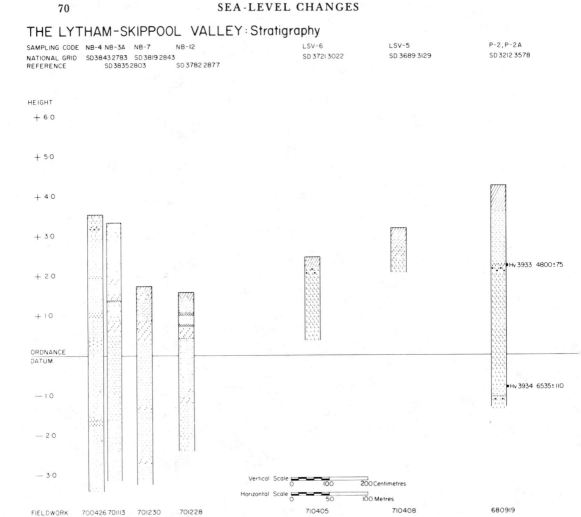

FIELDWORK   700426 701113   701230   701228   710405   710408   680919

FIG. 21. The stratigraphy at fifteen sampling sites in the Lytham-Skippool valley.

			towards the base and including some *Phragmites*. Upper contact within 1 cm.
2.	-1.07 to -1.10	532 to 535	As4 Lf+ Th[1] *(Phra.)*+ Buff clay with ferruginous partings and some *Phragmites*.
1	-1.10 to -1.33	535 to 558	As3 Ag1 Th[4] + Fine, silty, blue-grey clay with some rootlets. Impenetrable below 558: coring chamber smashed.

   The pollen diagram from Peel-2, is displayed in Fig. 22. The following Local Pollen Assemblage Zones are recognised:

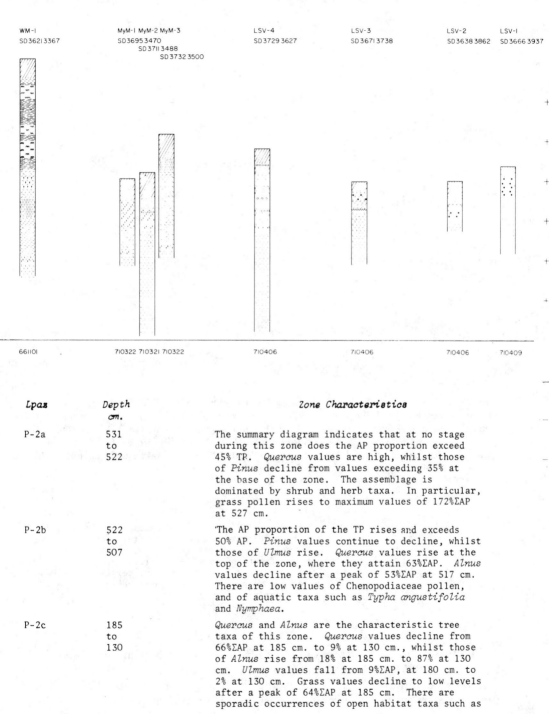

| WM-I SD36213367 | MyM-1 MyM-2 MyM-3 SD36953470 SD37113488 SD37323500 | LSV-4 SD37293627 | LSV-3 SD36713738 | LSV-2 SD36383862 | LSV-I SD36663937 |

661101   710322 710321 710322   710406   710406   710406   710409

Lpaz	Depth cm.	Zone Characteristics
P-2a	531 to 522	The summary diagram indicates that at no stage during this zone does the AP proportion exceed 45% TP. *Quercus* values are high, whilst those of *Pinus* decline from values exceeding 35% at the base of the zone. The assemblage is dominated by shrub and herb taxa. In particular, grass pollen rises to maximum values of 172%ΣAP at 527 cm.
P-2b	522 to 507	The AP proportion of the TP rises and exceeds 50% AP. *Pinus* values continue to decline, whilst those of *Ulmus* rise. *Quercus* values rise at the top of the zone, where they attain 63%ΣAP. *Alnus* values decline after a peak of 53%ΣAP at 517 cm. There are low values of Chenopodiaceae pollen, and of aquatic taxa such as *Typha angustifolia* and *Nymphaea*.
P-2c	185 to 130	*Quercus* and *Alnus* are the characteristic tree taxa of this zone. *Quercus* values decline from 66%ΣAP at 185 cm. to 9% at 130 cm., whilst those of *Alnus* rise from 18% at 185 cm. to 87% at 130 cm. *Ulmus* values fall from 9%ΣAP, at 180 cm. to 2% at 130 cm. Grass values decline to low levels after a peak of 64%ΣAP at 185 cm. There are sporadic occurrences of open habitat taxa such as

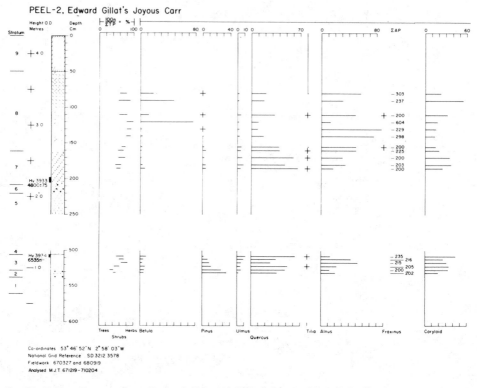

FIG. 22. Pollen diagram from Peel − 2, Edward Gillat's Joyous Carr, P-2. The frequency of each taxon at successive levels has been calculated as a percentage of the sum of the arboreal pollen, and the calculating formula is shown. The AP sum comprises *Betula*, *Pinus*, *Ulmus*, *Quercus*, *Tilia*, *Alnus* and *Fraxinus*. The two samples for radiocarbon dating were taken from a piston core sample. Samples for pollen analysis from 50 to 190cm were taken with a Hiller, whilst those from 500 to 530cm derived from the piston corer.

		Filipendula, *Taraxacum*-type, *Centaurea*, *Galium* and *Pteridium*.
P-2d	130 to 80	*Betula* rises to high frequencies at the base of the zone, but declines thereafter, notwithstanding a collapse of values at 110 cm. and the partial replacement of *Betula* by *Quercus* and *Alnus*. There are low, fluctuating but persistent values of *Salix*, and sporadic occurrences of open habitat taxa throughout the zone.

The two radiocarbon dates of 6535±110 from 506 to 510 cm and 4800±75 from 198 to 205 cm allow the Flandrian II/III Chronozone boundary to be interpolated tentatively at some point in Strata 4, 5 and 6. The pollen diagram does not permit greater precision, for biogenic deposition at the site

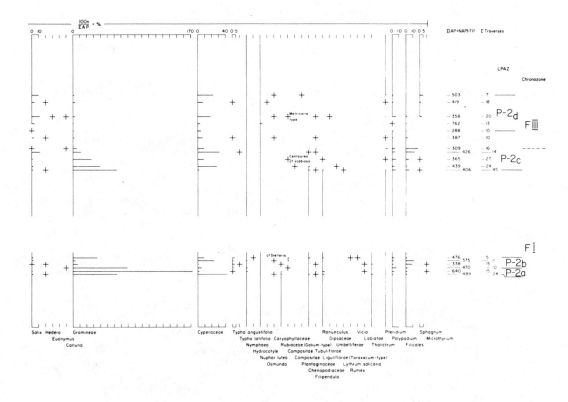

had ended before the culminating stage of FII and did not recommence until the early part of FIII, at which time *Ulmus* values are low, and ruderal taxa, albeit sporadic, are present.

Lytham Moss

The south-western part of Lytham Moss is shown on Fig. 24. The seaward limits of the moss are taken as Heyhouses Lane and Moss Edge Lane, which follows the +7.5 metre contour northwards from Heyhouses Lane. This margin of the moss has been overblown by sand. To the east, the moss is limited by rising till, which forms a broken ridge at Peel, Lower Ballam and Lytham Hall Park. Locally, as at Midgeland, low till hills emerge. To the north, beyond Division Lane, Lytham Moss runs into Great Marton Moss.

Most of the peat on Lytham Moss has been cut for fuel or burnt. Only veneers of peat survive at the margins of the moss (Wray and Cope 1948). At the centre of the moss, ground altitudes are +3.0 metres, and rise towards the margins of the moss.

The moss is underlaid by blue-grey, silty clays, which are exposed at the surface east of Queensway, and occur at altitudes across the moss from +2.6 to

LYTHAM   MOSS — I THOMAS   SALTHOUSE'S   LONGFIELD

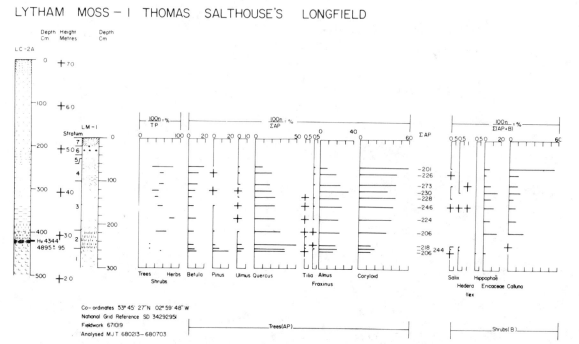

FIG. 23. Pollen diagram from Lytham Moss-1, at the south-western limit of the moss, where it has been overblown by sand. The frequencies of each taxon for successive levels through the deposit derive from the calculating formula shown for each group; AP, B, C, D, E. The upper 500cm from an adjacent sampling site, LC-2A, are also given so that the radiocarbon date of 4895±95 can be shown in its correct stratigraphic position in relation to LM-1.

+3.6 metres O.D. (Fig. 25). The silts and clays are 9.45 metres thick at LM-15 (Fig. 25), and to the east of Kite Hall Wood (SD 354306) a shell-bearing silty clay 7.16 metres thick has been reported: the shells included *Cerastoderma edule* Linné and *Tellina tenuis* da Costa.

Lytham Moss—1, LM-1 Thomas Salthouse's Long Field

At the very margin of the moss, where peat has been protected by the over-burden of sand, it was possible to obtain samples for pollen analysis. Samples were taken with a Hiller-type peat sampler and the following stratigraphic succession was noted:

Stratum	Height O.D. metres	Depth cm.	Description
7	+5.27 to +5.09	000 to 018	Str. conf. Sandy soil.

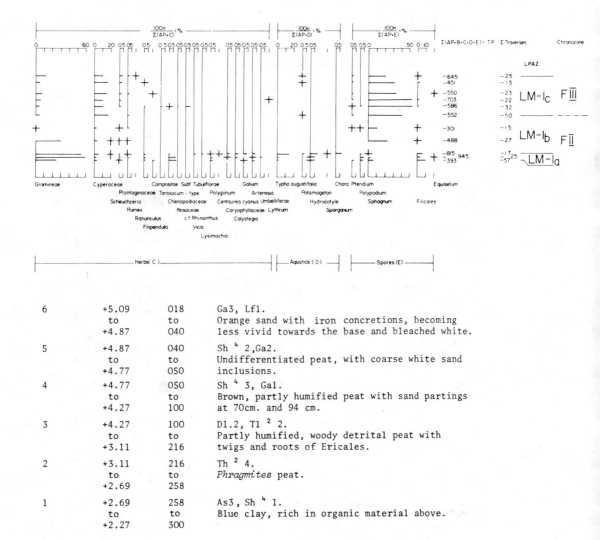

6	+5.09 to +4.87	018 to 040	Ga3, Lf1. Orange sand with iron concretions, becoming less vivid towards the base and bleached white.
5	+4.87 to +4.77	040 to 050	Sh⁴2,Ga2. Undifferentiated peat, with coarse white sand inclusions.
4	+4.77 to +4.27	050 to 100	Sh⁴3, Ga1. Brown, partly humified peat with sand partings at 70cm. and 94 cm.
3	+4.27 to +3.11	100 to 216	Dl.2, Tl²2. Partly humified, woody detrital peat with twigs and roots of Ericales.
2	+3.11 to +2.69	216 to 258	Th²4. *Phragmites* peat.
1	+2.69 to +2.27	258 to 300	As3, Sh⁴1. Blue clay, rich in organic material above.

Two stratigraphic columns are displayed with the pollen diagram (Fig. 23), one from the sampling site, the other from an adjacent site (LC-2A, SD 33942946) some 340 metres west. The altitudinal levels of similar strata at both sampling sites are similar: the boundary between Strata 2 and 1 in LM-1 occurs at +2.69 metres O.D. whereas that between Strata 14 and 13 in LC-2A

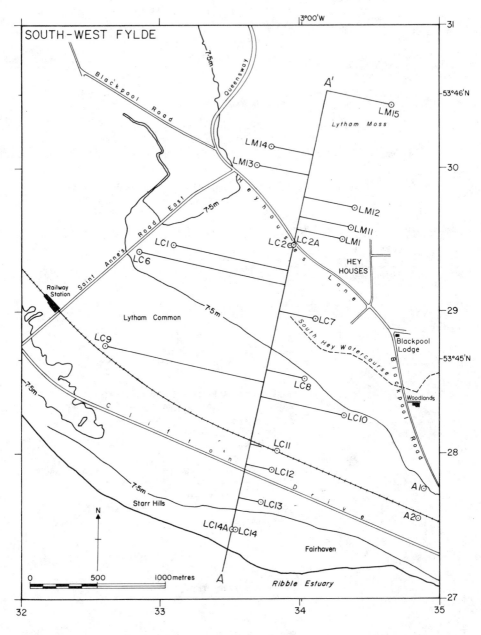

FIG. 24. A map of the south-west Fylde to show the location of sampling sites across Lytham Common and Lytham Moss, projected onto an artificial line A-A . The area is now more or less continuously built up, constituting part of Lytham St. Annes. The following sampling codes have been used: LM-Lytham Moss series, LC-Lytham Common series, A-Ansdell series.

is at +2.84 metres O.D.; between 5 and 6, and 17 and 18 it is +4.87 and +4.94 metres respectively. The successional development is also similar.

The samples from LM-1 were too small for radiocarbon dating and were liable to contamination during the sampling procedure, whereas samples from LC-2A derived from a U4 core. Three Local Pollen Assemblage Zones have been recognised at LM-1 (Fig. 23):

LPAZ	Depth cm.	Zone Characteristics
Lm-1a	260 to 245	The AP proportion falls suddenly, and low values not exceeding 27% TP are maintained. AP values are dominated by *Quercus*, which rises to 50%ΣAP at the top of the zone. *Pinus* values fall from 20% to 30%ΣAP, and those of *Ulmus* are sustained at levels greater than 4%ΣAP. Grass frequencies are high, rising from 25% to more than 50%Σ (AP+C). Chenopodiaceae values decline from 10% to 1%Σ(AP+C) at 260 and 245 cm. respectively. Both *Typha* and *Hydrocotyle* values rise towards the top of the zone.
LM-1b	245 to 155	The AP component of the TP rises and is dominated by *Quercus* and *Corylus*. Through the zone the former declines whilst the latter rises. Both *Pinus* and *Ulmus* become discontinuous. The proportion of shrubs increases, and of the dwarf shrubs, *Calluna* values rise and then fall. Grass values become discontinuous. Sphagna begin with modest frequencies and then decline.
LM-1c	155 to 65	*Corylus* maintains its pre-eminence among the AP and *Quercus* continues to fall. *Betula* values rise. Frequencies of shrubs increase at the top of the zone, almost entirely the result of *Calluna,* which at 80 cm. accounts for 12%Σ(AP+B) and at 65 cm. 56%.

The whole assemblage is assigned to Chronozone FIII. The radiocarbon date of 4895±95 from LC 2A places the basal biogenic deposit well within the chronozone limits established by Hibbert *et al.* (1971) from Red Moss, but the behaviour of both *Pinus* and *Ulmus*, compared to their behaviour at LC-1 would indicate either FII for the basal samples or a long transitional zone between FII and FIII at this sampling site, or a depositional hiatus here. There is, indeed, some evidence for the latter in the behaviour of the AP frequencies and the sharp boundary between Strata 1 and 2.

Lytham Common

The area comprising Lytham Common (Fig. 24) lies south of Heyhouses Lane and west of Lytham Hall Park (Blackpool Road). It is bounded in the south and west by the Ribble Estuary and the Irish Sea. Most of the surface comprises blown sand. The area is now built up and constitutes part of Lytham St. Annes.

The greater part of the Common lies above + 7.0 metres O.D., but the land

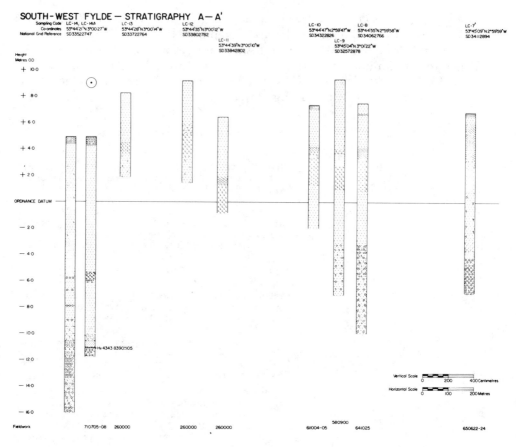

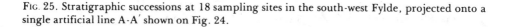

Fɪɢ. 25. Stratigraphic successions at 18 sampling sites in the south-west Fylde, projected onto a single artificial line A-A′ shown on Fig. 24.

falls imperceptibly northwards towards Lytham Moss, into which it was formerly drained by the South Hey Watercourse, and more steeply southwards and westwards towards the sea, beyond the belt of coastal sand dunes.

The mantle of blown sand obscures the topography which derived from weathered Devensian drift and the Flandrian sediments.

The cross-section displayed in Fig. 25 shows the undulating till surface: at LC-2, the till surface occurs at −10.6 metres O.D., while at LC-7 the till surface is at −4.2 metres O.D., and at LC-10 it has risen to −1.77 metres O.D. On Lytham Common, the lowest level at which the till surface was recorded occurs at LC-14, the Starr Hills, where the transition from till to a peaty sand occurs at −11.2 metres O.D.

On this undulating till surface, peats, silts, clays and sands of Flandrian age

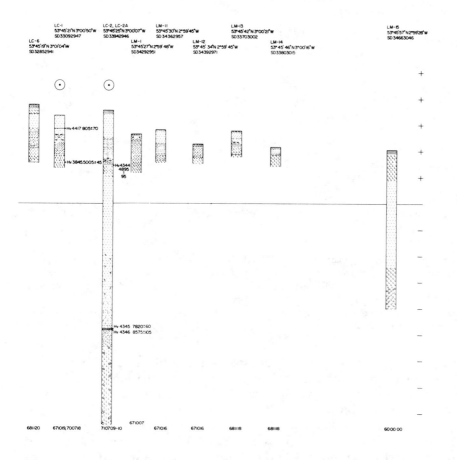

have accumulated. The greatest thickness of Flandrian sediments was recorded at Heyhouses Lane (LC-2A), where more than 17 metres have been proved.

A lower peat has been recorded at two sampling sites (LC-2 and LC-14) lying on the weathered surface of the till, and overlaid in turn by marine clays and silts, which are locally rich in valves of *Scrobicularia plana* da Costa. The sedimentation of these clays and silts is replaced either by peat from LC-11 landward or by blown sand from LC-12 seaward. The peat feathers both landward on to Lytham Moss, where it has been cut, and seawards towards the Starr Hills and the present coast. The greatest surviving thickness of peat occurs in Godwin's Sandpit (SD 32853084), where it exceeds 380 cm.

In every case, the peat is overlaid by blown sand of varying thickness: at LC-4 (SD 33212956) it is 0.45 metres thick, and at LC-2, 2.1 metres thick. On the Royal Lytham and St. Annes Golf Course it reaches 4.5 metres in thickness (LC-9).

THE STARR HILLS, LYTHAM

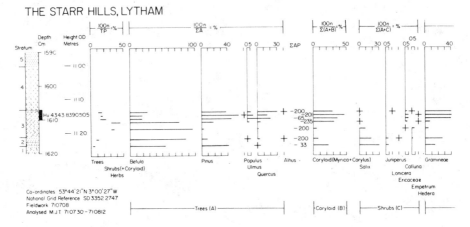

FIG. 26. Pollen diagram from the Starr Hills, Lytham, LC-14A. The frequencies of each taxon at successive levels through the biogenic deposit derive from the calculating formula shown for each group. The following groups are recognised: AP, Tree taxa: B, *Corylus* and *Myrica*, which are not distinguished, are grouped into a Coryloid division: C, Shrub taxa: D, Herb taxa: E, Aquatic taxa: F, Pteridophyte and Bryophyte taxa: G, Varia, comprising fungal spores, algal communities and others.

Where the sand thickness exceeds one metre, it is differentiated into a lower and upper stage, separated by a palaeosol or an accumulation of sandy peat. There may be additional layers locally. At LC-6, there is a soil horizon at 55–56 cm and peaty horizons at 75–7 cm and 185–200 cm before woody, detrital peat is reached at 220 cm.

The belt of sand ranges in width from 0.5 to 3.0 km. To the north and east it has overblown Lytham Moss, and the alternating sand and peat strata at the sampling sites indicate the changing nature of the ecotone between the Starr Hills' sand dunes and the raised mosses to landward.

The artificial stratigraphic line A-A' (Fig. 25) across Lytham Common is based upon a collection of borings carried out between 1926 and 1971. The location of the sampling sites is shown on Fig. 24. The stratigraphic successions at LC-1 and LC-6 were proved by a Russian-type sampler, whilst those at LC-2A and LC-14A were proved by a shell and auger drill. Samples for pollen analysis and radiocarbon dating were obtained from LC-1 by the Russian sampler, and from LC-2A and LC-14A by a series of U4 core samples.

The Starr Hills, Lytham LC-14A

In 1967, a series of borings proved the existence of a lower peat 15.54 to 16.76 metres from the ground surface. In 1971, another boring was put down in the same dune slack, a little to the west of the original site. The following strati-

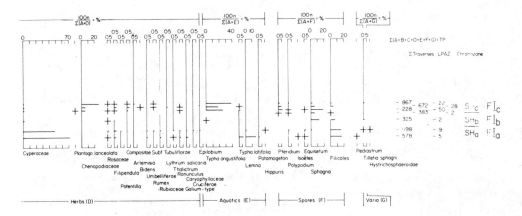

graphic succession was recorded from the U4 cores:

Stratum	Height O.D. metres	Depth cm.	Description
4	-11.00 to -11.13	1594 to 1607	Ag4 nig. 2, strf. 0, elas. 0, sicc. 2 Fine, grey clay.
3	-11.13 to -11.21	1607 to 1615	DL3 anth. 1 nig. 4, strf.1, elas. 0, sicc. 3, lim. sup. 4 Compact, well-humified, woody, detrital peat. Dense lense of charcoal at 1607, and a sharp contact above. Transition to
2	-11.21 to -11.24	1615 to 1618	Ag3 Sh^4 1 nig. 3, strf. 0, elas. 0, sicc. 3, lim. sup. 0 Organic sand.
1	-11.24 to -11.29	1618 to 1623	As3, Gs1, Gg+ nig. 2, strf. 1, elas. 0, sicc. 3, lim. sup. 0 Hard, tenacious pink clay, with angular stones and gravel

Two U4 core samples were taken from 1570 to 1618, and samples for pollen analysis and radiocarbon dating were taken from these cores. The peat layer was extremely hard and compact. The sampling interval varied from one to three centimetres. The pollen diagram is displayed as Fig. 26. Three Local Pollen Assemblage Zones are recognised:

Lpaz	Depth cm.	Zone Description
SH$_a$	1620 to 1618	The assemblage is dominated by non-arboreal pollen to the extent of 92% of the total pollen. The herbaceous taxa are particularly rich at this level and although Cyperaceae values account for 72% of the Σ(AP+B+C+D) open habitat taxa such as *Galium*, *Filipendula* and *Caryophyllaceae* are well represented. There are low values of *Potamogeton* and isolated peaks of *Typha latifolia* and *Lemna*.
SH$_b$	1618 to 1611	There is a strong rise in the proportion of arboreal pollen throughout the zone, and this is contributed largely by a single taxon, *Betula*, which at 1613 cm. accounts for 96%ΣAP. Single grains of *Ulmus* and *Alnus* are recorded and there is the beginning of a continuous *Quercus* curve. *Salix* values decline throughout the zone, and then become discontinuous as do those of both grasses and sedges. Coryloid values rise suddenly at the top of the zone.
SH$_c$	1611 to 1608	The AP values fall considerably throughout the zone. The frequency of *Betula* collapses from 76% at 1611 to 18%ΣAP at 1610, and these lower values are maintained. *Pinus* values rise, whilst those of *Coryloid* are sustained at values greater than 25%Σ(AP+B).
		Only dwarf shrubs, *Calluna*, are represented. Grass values rise strongly, and values of 40%Σ(AP+D) are sustained. Of the aquatics *Typha angustifolia* is well represented, rising towards the top of the zone. This pattern is reflected by the *Chenopodiaceae*.

The three Local Pollen Assemblage Zones, SHa,b,c, can be assigned to the Regional Pollen Assemblage Zones a, b, c, at Red Moss with some confidence (Fig. 4). A sample of woody detrital peat with charcoal fragments was taken from the top of Stratum 3 and yielded a date of 8390±105. Pollen spectra from the levels at which the sample was taken for radiocarbon dating show high *Pinus* and *Corylus* frequencies: *Ulmus* frequencies are persistent and continuous, whilst those of *Quercus* are rising. *Alnus* is represented only by single grains. On this basis, the radiocarbon-dated sample can be referred to the latter part of RPAZc at Red Moss and the date thereby corroborated (Fig. 4).

Heyhouses Lane, St. Annes LC-2A

Another lower peat deposit was reported from the grounds of the Government Offices in St. Annes in 1967. The organic band proved at sampling site LC-2A was a little over 10 cm thick and at a much lower altitude than that recorded in 1967, when an organic band 45 cm thick, 1570 cm from the surface was recorded. From the organic band, two samples 5 cm thick were cut for radio-carbon dating, from 1670–1674 cm and 1680–1684 cm, and samples were taken at 2 cm intervals for pollen analysis.

The basal stratigraphic succession at the site described from the U4 cores is as follows:

Stratum	Height O.D. metres	Depth cm.	Description
7	-9.59 to -9.62	1666 to 1669	Ag3, Sh [4] 1. Soft, grey silt rich in organic macro-fossils.
6	-9.62 to -9.66	1669 to 1673	Ld [4] 3. Sh [4] 1. Dl.+. As+. Gyttja with clay partings, rich in macro-fossil including woody fragments.
5	-9.66 to -9.71	1673 to 1678	Th [3] 4. Hard, compact, monocot. peat.
4	-9.71 to -9.75	1678 to 1682	Ld [4] 3. Th [3] 1. Gyttja. Black. Occasional monocot. fragments. Sharp break at base.
3	-9.75 to -9.88	1682 to 1695	As4. Ga+. Fine, grey clay with few sandy partings.
2	-9.88 to -9.97	1695 to 1704	As3. Ga1. Coarse grey clay with sandy partings.
1	-9.97 to -10.01	1704 to 1708	As3. Ga1. Sh [4] +. Buff clay with discrete sandy, pink clay and organic partings.

Samples were taken for pollen analysis from Stratum 5, and the pollen diagram is given in Fig. 27. Two Local Pollen Assemblage Zones are recognised:

Lpaz	Depth cm.	Zone Characteristics
H1a	1684 to 1678	The assemblage is characterised by high and sustained values of *Betula* and *Corylus*. *Populus* values are low but continuous, whilst those of *Quercus* fluctuate, but exceed those of *Ulmus* and are also continuous. *Salix* values increase and reach a peak at the top of the zone. A range of aquatic taxa is represented, among which *Typha latifolia* is pre-eminent. There are low but persistent frequencies of *Pediastrum* communities. Fern spores sustain moderate frequencies.

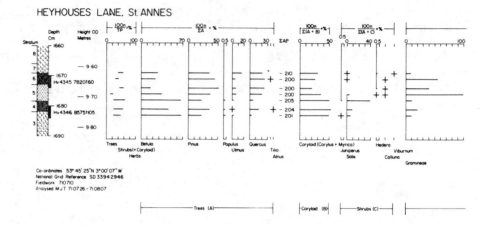

Fig. 27. Pollen diagram from Heyhouses Lane, St. Annes, LC-2A. The frequency of each taxon at successive levels through the biogenic deposit is based on the calculating formula shown for each group. Seven groups are recognised: AP, Tree taxa: B, Coryloid which comprises *Corylus* and *Myrica*: C, Shrub taxa: D, Herb taxa: E, Aquatic taxa: F, Pteridophyte and Bryophyte taxa: G, Varia comprising algal communities and unidentified taxa.

| H1b | 1678 to 1668 | *Betula* frequencies fall and are replaced by high *Pinus* values. *Quercus* values exceed 20%ΣAP, and although *Ulmus* is continuous its values never exceed those of *Quercus*. *Corylus* values fall, whilst those of *Salix* collapse and become discontinuous. There is a spectacular rise in the frequencies of grass pollen and at 1676 accounts for 99%Σ(AP+D). Of the aquatics only *Typha angustifolia* is well represented and peaks during the zone. The frequency of Chenopodiaceous pollen rises towards the end of the zone. |

On the basis of the two radiocarbon dates (Table 1 and Fig. 27), the whole assemblage can be assigned to Chronozone FI, which is supported by the concurrence of the two Local Pollen Assemblage Zones, HLa, b with Regional Pollen Assemblage Zones c and d at Red Moss.

Lytham Common—1. LC-1 Thomas Gillat's Colley Hey

The presence of marine silt and clay overlain by peat beneath an overburden of blown sand was proved at LC-1 in 1967, using a Hiller-type peat sampler. Samples for pollen analysis and radiocarbon dating were taken with a Russian-type sampler in 1968 and 1970. The site, like that at LC-6, lies below the +7.5 metre contour, in an area where the sand thickness attenuates at the head of a small shallow valley which drained east to the South Hey Watercourse (Fig. 24).

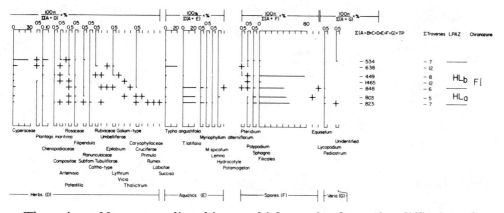

There is a 30 cm sampling hiatus which results from the difficulty of obtaining samples from the peat-sand interface. The stratigraphic succession at the site derives from the two latter sampling attempts which proved successful: the two parts of the succession shown diagrammatically in Fig. 28 above and below the unsampled stratum are not exactly coincidental, but are given as if part of a single core.

Stratum	Height O.D. metres	Depth cm.	Description
15	+6.67 to +6.49	000 to 018	Str. conf. Sandy, surface soil.
14	+6.49 to +6.24	018 to 043	Ga3. Sh[4] 1. Organic sand.
13	+6.24 to +6.16	043 to 051	Ga4. Yellow sand.
12	+6.16 to +6.13	051 to 054	Sh[4] 3. Ga1. Sandy peat.
11	+6.13 to +5.78	054 to 089	Gs3. Lf1. White, coarse sand with ferruginous partings.
10	+5.78 to +5.75	089 to 092	Th[2] 4. Ga+. Monocot. peat, partly humified, with a slight sand fraction.
9	+5.75 to +5.73	092 to 094	Sh[4] 2. Ga2. Sandy peat.
8	+5.73 to +5.65	094 to 102	Ga4. White to buff sand.

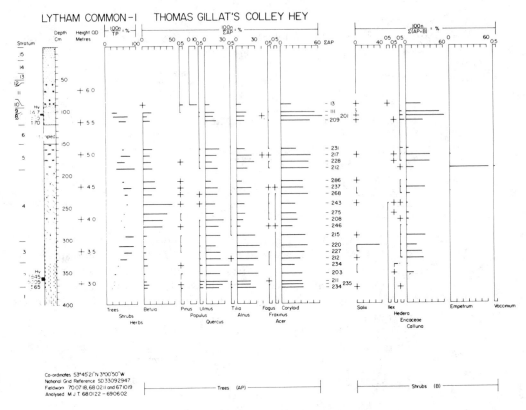

FIG. 28. Pollen diagram from Lytham Common—1, Thomas Gillat's Colley Hey LC-1. The upper part of the diagram from 0 to 120cm was sampled by a coring device in 1968, while the lower part from 150 to 400cm was sampled with a Hiller-type peat sampler in 1967. The samples for radiocarbon dating were obtained in 1970 using a Russian-type sampler.

The frequencies of each taxon at successive levels through the biogenic deposit derive from the calculating formula shown for each group. The following 5 groups are recognised: AP, Tree taxa including Coryloid; B, Shrub taxa: C, Herb taxa: D, Aquatic taxa: E, Pteriodophyte and bryophyte taxa.

7	+5.65 to +5.50	102 to 117	Th ^{2} 4. Ga+. Partly humified monocot. peat, with a slight sand fraction.
6	+5.50 to +5.17	117 to 150	Unsampled.
5	+5.17 to +4.77	150 to 190	Th ^{3} 3. Th ^{2} . (vagi.)1. Black amorphous peat, with some *Eriophorum* throughout.
4	+4.77 to +3.67	190 to 300	D1.3. Th ^{0} *(Meny.)*1. Partly humified peat, with twigs of *Alnus* and *Betula* throughout, and rich in *Menyanthes* seeds.

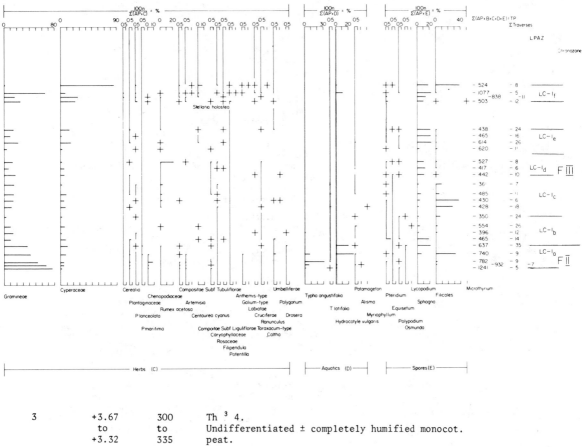

3	+3.67 to +3.32	300 to 335	Th ^{3} 4. Undifferentiated ± completely humified monocot. peat.
2	+3.32 to +2.94	335 to 373	Th ^{3} 3. Th ^{1} (Phra.)1. Partly humified monocot. peat with *Phragmites*, particularly at 335,to 342, 347 to 350 and 355 to 374.
1	+2.94 to +2.67	373 to 400	As3. Th ^{1} (Phra.)1. Blue, grey clay, rich in *Phragmites* rhizomes above.

From the pollen diagram (Fig. 28), six Local Pollen Assemblage Zones are recognised:

LPAZ	Depth cm.	Zone Characteristics
LC-1a	380 to 340	The assemblage is dominated by non-arboreal pollen. The AP proportion of the TP rises from 19% at 375 to 36% at 340 cm. Gramineae values are high but decline throughout the zone. The AP values are dominated by *Quercus, Alnus*

and *Corylus*. *Ulmus* values sustain low frequencies of 5%ΣAP, or more. *Typha angustifolia* peaks during the zone and the frequency of *Hydrocotyle* reaches a maximum of 28%Σ(AP+D) at 353 cm. Sphagna also rises and there are low values of *Drosera*.

LC-1b   340
         to
         290

The zone is characterised by *Alnus*, *Quercus* and *Corylus*, whose representation changes only slightly through the zone. *Salix* values rise and reach a maximum of 37%Σ(AP+B) at 310 cm. Dwarf shrub frequencies, especially those of *Calluna* also rise and are sustained throughout the zone at a level exceeding 17%Σ(AP+B).
Ulmus values never exceed 2% AP, and the frequency falls by more than half at the boundary. The first open habitat taxa, such as *Plantago lanceolata*, *Rumex acetosa*, *Taraxacum*-type and *Pteridium* occur in this zone.

LC-1c   290
         to
         230

Betula values rise, and pollen of this taxon dominates the AP. A fall in *Betula* values at the top of the zone is compensated by a rise of *Corylus*. Fern spores rise in frequency and reach a maximum value of 37%Σ(AP+E) at 270 cm.

LC-1d   230
         to
         200

Corylus values dominate the zone assemblage together with the other AP taxa. Dwarf shrub taxa again rise to peak values of 31%Σ(AP+B), and at the top of the zone *Rumex* values exceed 20%Σ (AP+C). There are low values of other open habitat taxa, especially *Plantago* and *Pteridium*.

LC-1e   200
         to
         150

AP values rise throughout the zone, and are characterised by *Corylus* as in the previous zone. The diagnostic feature of the zone is the large proportion of shrub taxa, which proportion decreases towards the top of the zone. In particular, the dwarf shrubs dominate the shrub assemblage: *Calluna* rises to 38%Σ(AP+B) at 170 cm. and *Empetrum* to 63%Σ(AP+B) at 190 cm. Sphagna values rise and sustain levels exceeding 20%Σ(AP+E) at the top of the zone.

LC-1f   120
         to
         90

Although the AP is dominated once again by *Corylus*, the proportion of AP declines rapidly throughout the zone, so that at 116 cm. the proportion is 42% TP, at 91 cm. it has fallen to 2% TP. The proportion of shrubs, especially dwarf shrubs also declines after high values (48% TP) are attained at 109 cm. contributed largely by *Calluna*. Grass values rise to a peak of 64% (AP+C) at 103 cm. and then collapse, whilst those of Cyperaceae peak at the top of the zone: 85% (AP+C) at 91 cm. There are low but persistent frequencies of cereal pollen, and an increase in the number and frequency of open habitat taxa, such as *Plantago lanceolata*, *Rumex*, *Artemisia*, *Centaurea cyanus* and members of the Rosaceae family.
Pteridium values are low and decline, whilst those of the Filicales rise.

On the basis of the radiocarbon dates, two chronozones, FII and FIII, are recognised, and the exact boundary has been drawn at 340 cm on the basis of the change in pollen frequencies of certain plant taxa. Although *Pinus* values have already become discontinuous, *Ulmus* values are low, but sustained at levels greater than 5 per cent AP. Open habitat taxa, such as *Plantago maritima*, *Artemisia* and other members of the Compositae, are recorded throughout LPAZ LC-1a, but their presence is interpreted as a response to marine conditions near the site. In the succeeding LPAZ, LC-1b, taxa associated with forest clearance, such as *P. lanceolata* and *Pteridium*, are recorded, and not only do they characterise the pollen assemblage at this level, they also help to specify the boundary between the chronozones.

Ansdell—1. A-1 Rossall Road

Two sampling sites, Ansdell-1 and Ansdell-2, have been examined in the south-east corner of Lytham Common (Fig. 24) in an attempt to clarify the stratigraphic successions in the south-western quadrant of Lytham Hall Park.

The area is mantled by blown sand, and the stratigraphy can only be observed in temporary sections and examined from cores. The thickness of the over-burden of sand prevents the use of hand-sampling equipment.

In Rossall Road, Ansdell, a series of borings was put down in 1970 and material from one borehole obtained for further analysis.

Three levels of the U4 core were analysed for pollen, but only one level had sufficient pollen and spores to allow a reasonable calculating sum to be used. The pollen spectrum from Ansdell-1 is displayed in Fig. 29. A radiocarbon assay on the peat yielded a date of 1370±85.

The assemblage is dominated by non-arboreal pollen, in particular by grass pollen, of which a very small proportion $(0.3\%\Sigma(AP+B))$ can be assigned to the Cereals. Notwithstanding the pre-eminence of grasses, there are low frequencies of open habitat taxa such as *Galium*-type and *Rumex*. Of the tree taxa, *Quercus* and *Betula* are the most important, but there are low frequencies of both *Fagus* and *Carpinus*.

Ansdell—2. A-2 Ansdell Railway Siding

The second site at Ansdell was chosen because of its accessibility and proximity to the present coast. Two U4 cores were obtained from this site, and the stratigraphic succession described below was taken from them:

Stratum	*Height O.D. metres*	*Depth cm.*	*Description*
5	+8.38 to +4.42	000 to 396	Str. conf. Ga4. Sandy Soil.
4	+4.42 to +4.23	396 to 415	Sh⁴ 4. Str. conf. Undifferentiated peat. Stratum disturbed by cutting shoe and sample lost.

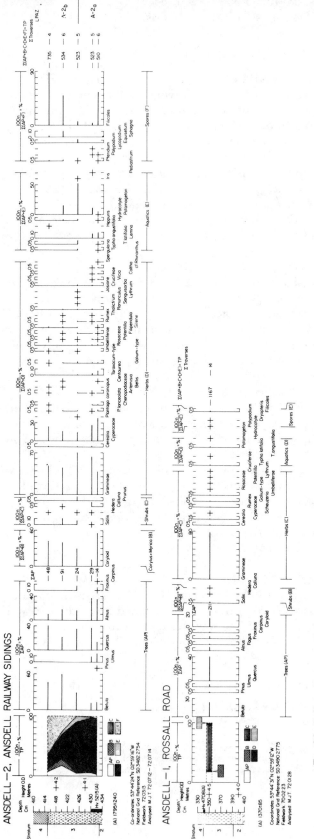

FIG. 29. Pollen diagrams from Ansdell. Ansdell − 1, Rossall Road and Ansdell − 2 Ansdell Railway sidings. At Ansdell − 1, five groups are recognised: AP, Tree taxa including Coryloid: B, Shrub taxa: C, Herb taxa: D, Aquatic taxa: and E, Pteriodophyte taxa. At Ansdell − 2, six groups are recognised: AP, Tree taxa: B, Coryloid: C, Shrub taxa: D, Herb taxa: E, Aquatic taxa, F, Pteridophyte and Bryophyte taxa. In both diagrams, percentage frequencies for taxa in each group are derived from the calculating formula given for each group.

3	+4.23 to +4.05	415 to 433	Dl.1. Th [2] 3. Th [1] *(Phra.)*+. Gs+. nig. 2, strf. 1, elas. 0, sicc. 2, lim. sup. 0 Woody, detrital peat passing down into a partly humified (H2) dry, compact, hard monocot. peat with *Phragmites*. A coarse sand fraction throughout. Deposit light brown. Sand fraction increases towards base.
2	+4.05 to +3.96	433 to 442	Ag2. Ga2. Th [1] +. As+. nig. 1, strf. 0, elas. 0, sicc. 2, lim. sup. 0 Medium, sandy silt, with monocot. roots and clay partings.
1	+3.96 to +3.50	442 to 488	Ag4. nig. 1, strf. 0, elas. 0, sicc. 2, Light, buff silt. Homogeneous. Disintegrates when dry.

Stratum 3 was analysed at 4 cm intervals, and the resulting pollen diagram
is displayed in Fig. 29. Two Local Pollen Assemblage Zones are recognised:

LPAZ	Depth cm.	Zone Characteristics
A-2a̅	432 to 425	The pollen assemblage is dominated by non-arboreal pollen. At no stage do the AP values exceed 5% TP, and this is maintained throughout the zone. The NAP proportion is dominated by herb taxa, and in particular by Gramineae and Cyperaceae. Cereal frequencies are low but persistent and there are low values of open habitat taxa such as *Plantago lanceolata*, *Galium*-type, *Rumex*, *Jasione*, *Vicia* and composites, for example *Artemisia*, *Centaurea* and *Taraxacum*-type. Aquatic taxa are conspicuous especially at the base of the zone, but *Potamogeton* values rise to a maximum of 53%Σ(AP+F) at 425cm.
A-2b	425 to 415	There is a spectacular rise in the proportion of spores, at the expense of herb taxa particularly, towards the top of the zone. AP values rise slightly to a peak of 17% TP at 420 cm., the most important component increase being in *Betula*. Fern spores rise from 7%Σ(AP+F) at 425 cm. 89% at 415 cm. The proportion of aquatic taxa decreases, largely the result of the decline of *Potamogeton*. The number and frequency of herb taxa is considerably reduced in this zone compared to A-2a.

A 4 cm thick sample was cut between 431 and 434 cm from the base of
Stratum 3 and the top of Stratum 4 for a radiocarbon assay, and yielded a
date of 1795±240. The pollen assemblage does not help to corroborate this
date, except in that the persistent Cereal pollen frequencies, the rich ruderal
assemblage and the record of *Carpinus* indicate a Flandrian III age.

Lytham Hall Park

Lytham Hall Park is a subdivision of Lytham Common, and is treated as a separate entity because development within the Park from 1968 onwards allowed a series of undisturbed samples to be taken from free-face excavations. The stratigraphic succession recorded in the Park is similar to that found elsewhere on the Common west of the Park boundary (see, for example, Lytham Common -1, and Ansdell -1 and -2).

Lytham Hall Park is taken to be that area enclosed by the circumvallating pebble walls in 1968 to the west, south and east of the Park. Its northern limits are defined by Liggard Brook.

Advance work for development in the Park involved the removal in whole or part of the Big Wood Plantations laid out by Thomas Clifton in the nineteenth century, and the setting out of sewerage and drainage lines. The borehole records together with stratigraphic descriptions from the excavations are the basis of the stratigraphic diagram displayed in Fig. 30.

The diagram shows a basin in the till, the lowest part of which is recorded in BH-24 at −5.46 metres OD. The basin rises both westwards and eastwards, where its edge becomes a landscape feature north of Lytham Hall. The basin is infilled with clays, silts, peats and sands. The contractors' records show a simple replacement of till by silty clay, peat and sand, but the detailed stratigraphic records from the excavations (sampling codes LHP-4, -5, -8 on Fig. 30) show that in the upper stages of the sedimentary sequence of silty clay, there is an alternating succession of clays and biogenic deposits. Further, the overburden of blown sand could be subdivided into an upper and lower unit, on the basis of a persistent palaeosol or peat bed intercalating the sand, and comparable to the successional sequence recorded elsewhere on Lytham Common, and along the western margin of Downholland Moss. The bed was best developed in a trench flanking LHP-8 (SD 34902858). Embedded within the peat were single valves of the Common Cockle *Cerastoderma edule* (Linné) and the mussel *Mytilus edulis* Linné. Samples were taken for pollen analysis, but successive levels through the deposits were extremely poor both in numbers and taxa, and a diagram could not be constructed: the only well-represented pollen taxon was *Taraxacum*-type.

Monolith samples were taken from LHP-4 and LHP-8 for radiocarbon dating. Earlier, samples were taken from LHP-1 and LHP-5 for pollen analysis: these sites were subsequently obliterated before undisturbed bulk samples could be taken.

Lytham Hall Park—1. LHP-1 Thomas Clifton's Plantation

Samples for pollen analysis were taken in 1968 from the west side of a free-face excavation at the south-western limit of the Park (Fig. 30). The upper part of the excavation was obscured by slumping and steel-sheet piling, but, in the lower part of the excavation, the following succession was recorded:

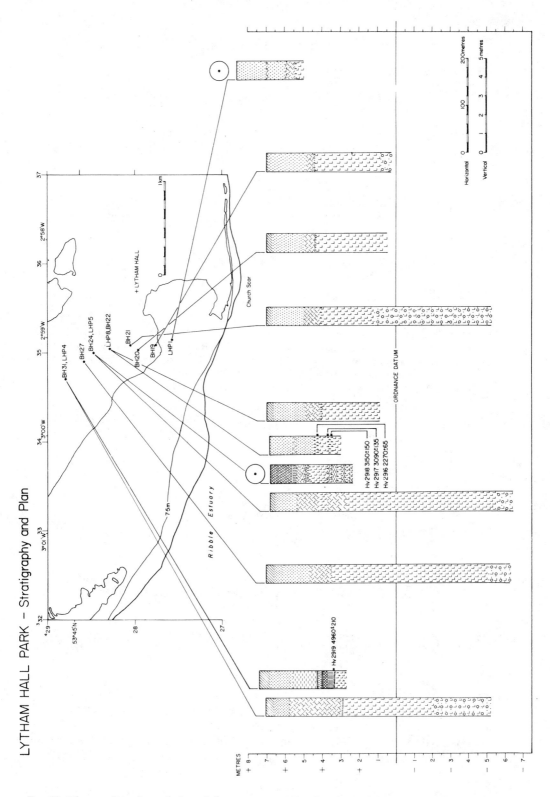

FIG. 30. The stratigraphy and plan of the western margin of Lytham Hall Park. Sampling codes prefixed by BH. were carried out by the Cementation Co. Ltd., and those prefixed by LHP were recorded from open-face excavations (LHP-4,-1) or from a Hiller-type peat sampler (LHP-5). Pollen analyses were carried out at LHP-1 and LHP-5 and are indicated by a circumscribed dot.

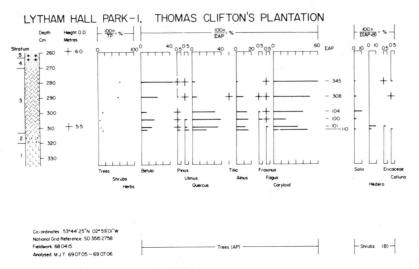

FIG. 31. Pollen diagram from Lytham Hall Park — 1, Thomas Clifton's Plantation, LHP-1. The percentage frequency of each taxon has been calculated for successive levels according to the formula for each group. Five groups are recognised: AP, Trees including Coryloid: B, Shrub taxa: C, Herb taxa: D, Aquatic taxa: E, Pteridophyte and Bryophyte taxa.

Stratum	Height O.D. metres.	Depth cm.	Description
6	+8.59 to +7.34	000 to 125	Ga4. Str. conf. Yellow sand, with many organic bands, obscured by slumping.
5	+7.34 to +5.93	125 to 266	Ga3. Sh^4 1. Grey, white sand below becoming yellow above with organic bands.
4	+5.93 to +5.89	266 to 270	Ga4. White sand.
3	+5.89 to +4.44	270 to 315	D1.3. Th^2 (Phra.)2. Highly compressed woody detrital peat, with monocotyledonous fragments, including *Phragmites*.
2	+4.44 to +5.39	315 to 320	As2. Th^2 (Phra.)2. Clay, rich in *Phragmites*.
1	+5.39 to	320+	As4. Tenacious, grey, blue clay.

Samples were taken every 10 cm through the biogenic deposit, Stratum 3, and the interval was increased from 5 cm to 2 cm as Stratum 2 was approached. In the pollen diagram (Fig. 31), two Local Pollen Assemblage Zones are recognised:

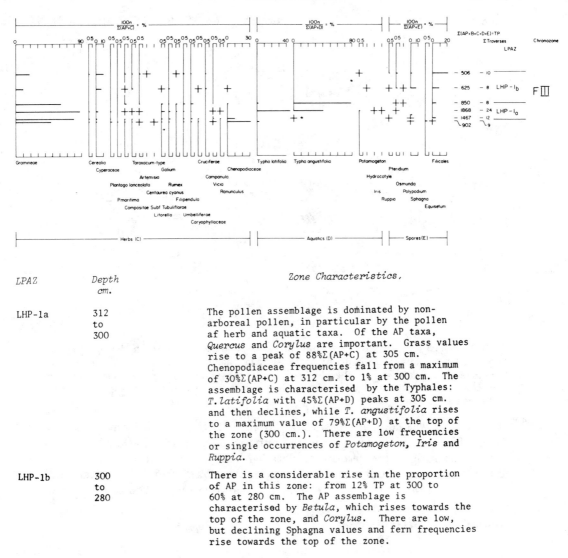

LPAZ	Depth cm.	Zone Characteristics.
LHP-1a	312 to 300	The pollen assemblage is dominated by non-arboreal pollen, in particular by the pollen af herb and aquatic taxa. Of the AP taxa, *Quercus* and *Corylus* are important. Grass values rise to a peak of 88%Σ(AP+C) at 305 cm. Chenopodiaceae frequencies fall from a maximum of 30%Σ(AP+C) at 312 cm. to 1% at 300 cm. The assemblage is characterised by the Typhales: *T. latifolia* with 45%Σ(AP+D) peaks at 305 cm. and then declines, while *T. angustifolia* rises to a maximum value of 79%Σ(AP+D) at the top of the zone (300 cm.). There are low frequencies or single occurrences of *Potamogeton*, *Iris* and *Ruppia*.
LHP-1b	300 to 280	There is a considerable rise in the proportion of AP in this zone: from 12% TP at 300 to 60% at 280 cm. The AP assemblage is characterised by *Betula*, which rises towards the top of the zone, and *Corylus*. There are low, but declining Sphagna values and fern frequencies rise towards the top of the zone.

The discontinuous record of *Ulmus*, the high *Corylus* frequencies, the presence of *Fagus*, and the unmistakable indicators of human activity close to the site, such as the presence of cereal pollen, *Plantago lanceolata*, *Taraxacum*-type and *Centaurea cyanus*, suggest a Flandrian III age for Stratum 3.

Lytham Hall Park—5. LHP-5 Francis Fox's South Hey Watercourse Meadow

Before development obscured the site, the northern bank of the South Hey

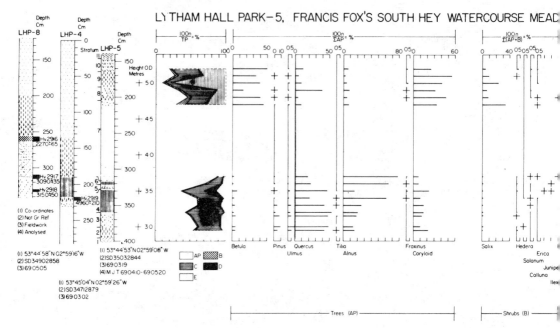

FIG. 32. A pollen diagram from Lytham Hall Park — 5, Francis Fox's South Hey Watercourse Meadow. Five groups are recognised: AP, Tree taxa, including Coryloid: B, Shrub taxa: C, Herb taxa: D, Aquatic taxa: E, Pteridophyte and Bryophyte taxa. The percentage frequencies at successive levels are derived from the calculating formulae given for the five groups and for the summary diagram.

Watercourse was cut back beside the drive to Lytham Hall from Blackpool Lodge (Fig. 24). The bed of the watercourse was composed of a woody detrital peat. The stratigraphic description detailed below, derives from a description of the stratigraphic succession in the watercourse bank from 0 to 145 cm and from a record taken with a Hiller-type peat sampler on the bed of the stream adjacent to the bank.

Stratum	Height O.D. metres	Depth	Description
11	+5.40 to +5.35	140 to 145	Gg (maj.) 4. Pebble bed.
10	+5.35 to +5.15	145 to 165	D1.2. Tl ^{2} 2. Partly humified, woody detrital peat, rich in twigs, roots and stumps of *Betula*.
9	+5.15 to +4.90	165 to 190	Th ^{2} 4. Th ^{1} *(Phra.)*+. Partly humified monocot. peat with *Phragmites* at 173 to 175 cm.
8	+4.90 to +4.75	190 to 205	Th ^{3} *(Phra.)*4. *Phragmites* peat, becoming humified towards the base.

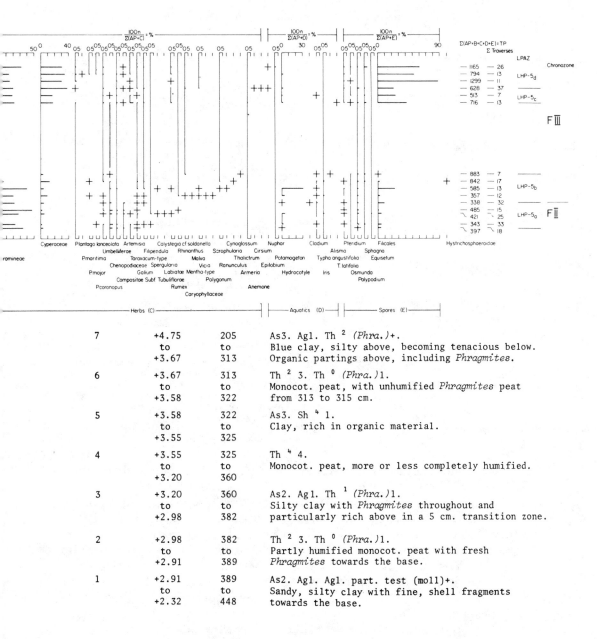

$\frac{100n}{\Sigma(AP+C)} = \%$ 　 $\frac{100n}{\Sigma(AP+D)} = \%$ 　 $\frac{100n}{\Sigma(AP+E)} = \%$

$\Sigma(AP+B+C+D+E) = TP$
Σ Traverses
LPAZ

— 1165	— 26	Chronozone
— 794	— 13	LHP-5d
— 1299	— 11	
— 628	— 37	
— 513	— 7	LHP-5c
— 716	— 13	

F III

— 883	— 7	
— 842	— 17	
— 585	— 13	LHP-5b
— 357	— 12	
— 338	— 32	
— 485	— 15	
— 421	— 25	LHP-5a	F II
— 343	— 33	
— 397	— 18	

Gramineae Cyperaceae Plantago lanceolata Artemisia Calystegia cf soldanella Cynoglossum Nuphar Cladium Pteridium Filicales Hystrichosphaeroidae
P maritima Umbelliferae Filipendula Rhinanthus Scrophularia Cirsium Alisma Sphagnum
P major Taraxacum-type Malva Thalictrum Potamogeton Typha angustifolia Equisetum
Chenopodiaceae Spergularia Vicia Ranunculus Epilobium T. latifolia
Galium Labiatae Mentha-type Armeria Hydrocotyle Iris Osmunda
Compositae Subf. Tubuliflorae Polygonum Polypodium
P coronopus Rumex Anemone
Caryophyllaceae

|— Herbs (C) —| |— Aquatics (D) —| |— Spores (E) —|

7	+4.75 to +3.67	205 to 313	As3. Ag1. Th ^{2} *(Phra.)*+. Blue clay, silty above, becoming tenacious below. Organic partings above, including *Phragmites*.
6	+3.67 to +3.58	313 to 322	Th ^{2} 3. Th ^{0} *(Phra.)*1. Monocot. peat, with unhumified *Phragmites* peat from 313 to 315 cm.
5	+3.58 to +3.55	322 to 325	As3. Sh ^{4} 1. Clay, rich in organic material.
4	+3.55 to +3.20	325 to 360	Th ^{4} 4. Monocot. peat, more or less completely humified.
3	+3.20 to +2.98	360 to 382	As2. Ag1. Th ^{1} *(Phra.)*1. Silty clay with *Phragmites* throughout and particularly rich above in a 5 cm. transition zone.
2	+2.98 to +2.91	382 to 389	Th ^{2} 3. Th ^{0} *(Phra.)*1. Partly humified monocot. peat with fresh *Phragmites* towards the base.
1	+2.91 to +2.32	389 to 448	As2. Ag1. Ag1. part. test (moll)+. Sandy, silty clay with fine, shell fragments towards the base.

Samples for pollen analysis were taken with a Hiller-type peat sampler, and the results are shown in Fig. 32. The sampling interval within the biogenic deposit ranged from 5 to 10 cm. The following four Local Pollen Assemblage Zones are recognised:

LPAZ	Depth cm.	Zone Characteristics
LHP-5a	385 to 350	The AP component accounts for more than 40% TP: of the arboreal pollen, *Quercus* values dominate the assemblage, and there are moderate values of *Alnus* and Coryloid. *Ulmus* values reach a maximum of 10%ΣAP at 380 cm., from which they decline through the zone.
LHP-5b	350 to 310	During this zone, AP values undergo a considerable reduction from 61% TP at 350 cm. to 36% at 330 cm. from which value they recover to 89% at the top of the zone, *Quercus* and *Alnus* again characterise the zone: *Quercus* values fall throughout the zone whilst those of *Alnus* rise suddenly and maintain high levels at the top of the zone. There is a rise in the frequency of aquatic taxa, and at 330 cm. *Hydrocotyle* rises to a peak value of 31%Σ (AP+D). The proportion of herb taxa falls throughout the zone. The a/b boundary is drawn at 350 cm. where *Ulmus* values fall from 6% to 1%ΣAP at 350 and 340 cm. respectively
LHP-5c	210 to 190	The AP values rise to a peak of 45% TP at 200 cm. and then decline through this and the subsequent zone. *Betula* values are high at first, and then decline, whilst *Quercus* behaves in complement. The c/d boundary is drawn at 290 cm. where AP values are declining and those of Pteridophytes and Bryophytes strongly increase.
LHP-5d	190 to 160	During this zone tree taxa undergo a considerable reduction to 9% TP at 180 cm. from which level they recover slightly later in the zone. Both *Betula* and Coryloid rise during the zone. *Salix* values continue to fall. Of the Pteridophytes, the ferns are pre-eminent and at 180 cm. account for 88%Σ(AP+E).

Two chronozones are indicated by referring Local Pollen Assemblage Zone LHP-5a to Regional Pollen Assemblage Zone e and LPAZ LHP-5b,c,d to RPAZf: the boundary between FII and FIII is drawn at 350 cm. At this level, *Ulmus* values are 6 per cent AP, having declined from a maximum of 10 per cent at 380 cm. Thereafter, they fall to 1 per cent at 340 cm. *Pinus* values are also low, and become discontinuous above 340 cm. The summary diagram (Fig. 32) shows a rise in the proportion of herb taxa, at the expense of tree taxa, especially at 350 cm, although there is a recovery at the chronozone boundary. The suggested FIII chronozone is corroborated by the three radio-carbon dates from biogenic deposits of an adjacent core: the stratigraphic succession and positions of the dated horizons are shown on Fig. 32.

The foregoing descriptions from sites on the Lancashire coast adjacent to the Ribble estuary exemplify the application of the methods described in Chapter 1 and the derivation and corroboration of information relevant to sea-level studies in humid, cool temperate latitudes.

CHAPTER 3

Coastal Sequences in North-West England

The data presented in Chapter 2 and considered elsewhere (Tooley 1969, 1974, 1976b, 1977a) are the basis for a description of temporal and spatial patterns of the coastal, sedimentary sequences recognised in north-west England. Patterns of marine transgressions are proposed for West Derby and Fylde in Amounderness, and a correlation is attempted between each area in north-west England. The coastal sequences from north-west England are shown schematically in Fig. 35 and this scheme serves as a basis for comparison with patterns elsewhere in North-west Europe (see Chapter 5).

LANCASHIRE: WEST DERBY

In West Derby, four main marine transgressions have been established on the basis of altitude (Table 3) and age. They are shown in Fig. 33 and schematically in Fig. 35. A fifth transgression can be inferred, the evidence for which is considered later.

The beginning of the first transgression is recorded at Long Lane, Formby, where biogenic sedimentation ended when estuarine silts began to accumulate there at an altitude of -10.21 metres O.D. On the Fylde, a marine transgression has been recorded at the Starr Hills on the present coast, at an altitude of -11.14 metres O.D. shortly after 8390±105 BP, and landward of this site at Heyhouses Lane, on Lytham Common, the same transgression is recorded at an altitude of -9.65 metres O.D. shortly after 7820±60 BP. The intermediate altitude of the transition to marine conditions at Long Lane, Formby, permits a tentative estimate of an early Flandrian Id age, that is about 8000 radiocarbon years, before the onset of the first transgression in West Derby. There is no stratigraphic evidence that this transgression was interrupted until it reached its landward limit between DM-11A and DM-12 (see Fig. 6) on Downholland Moss. This interpretation, however, is based on few borehole records to insufficient depth between Ranslett House, Formby, and DM-2 (Fig. 33), and records taken beneath the over-burden of blown sand may show a succession similar to the one recorded in the Fylde. At the eastern margin of Downholland Moss, at DM-11A, the beginning of marine conditions, the marine limit of the first transgression and its culminating stages, between 6980 and 6760 BP, are registered. At DM-11, the transgression is recorded by Stratum 3, which is a tenacious blue clay, rich in the rhizomes of *Phragmites* towards the top and base of the stratum. At DM-11A the transgression is recorded by a wedge of blue clay with black organic and yellow ferruginous partings, reminiscent of a salt marsh soil. The approach of marine conditions at DM-11 is presaged by the elevation of the freshwater table, favouring aquatic taxa such as *Potamogeton, Nymphaea, Typha*

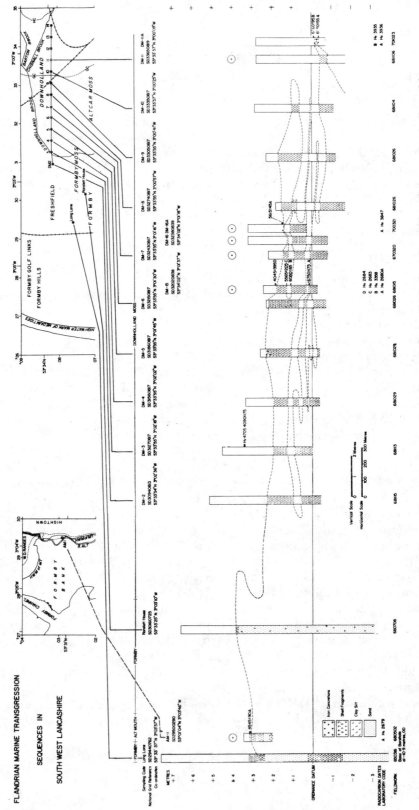

FIG. 33. Marine transgression sequences in Lancashire: West Derby. Sampling sites from which there are pollen diagrams are indicated by circumscribed points. The vertical exaggeration is x 125.

angustifolia and *Sparganium* and by the rise in the frequency of the pollen of Chenopodiaceae, Gramineae, *Artemisia* and *Armeria* immediately below the lithologic boundary. A progressive vegetation succession marks the end of marine conditions at DM-11: a *Phragmites*-rich monocotyledonous peat with some woody detritus, mainly of *Alnus*, is replaced entirely by a woody detrital peat with *Alnus* branches. The herbaceous pollen taxa show a reduction in the frequency of open habitat coastal taxa, such as the Chenopodiaceae, *Artemisia* and *Armeria* and their replacement by freshwater taxa such as *Nymphaea*, *Hydrocotyle*, *Typha angustifolia* and *Cladium mariscus*. At DM-11A, an open water lake mud began to accumulate, immediately after marine sedimentation ended.

At DM-15, 1.5 km seaward of DM-11A, the culminating stages of the first transgression have been recorded at an altitude of +0.04 m O.D. about 6750 radiocarbon years ago. As the lithologic boundary is approached, the marine sediments become progressively finer, until they are dominated by a clay fraction immediately below the boundary. The diatom flora changes from one dominated by marine forms, such as *Melosira sulcata* (Ehr.) and M. *westii* W. Smith in the silts at the base of the core to marine-brackish and freshwater and brackish forms, such as *Diploneis Smithii* (Breb.) Cleve and *Pinnularia viridis* (Nitzsch) Ehrenberg in the silty clay immediately below the lithologic boundary (see Appendix). Above the boundary and within the *Phragmites* peat, the proximity of marine conditions is shown by the low pollen frequencies of *Plantago maritima*, Chenopodiaceae and *Artemisia* and of freshwater plant communities by *Typha angustifolia*, *Cladium mariscus* and *Hippuris*.

In West Derby, the type area for both the transgressive and regressive stages of the first marine transgression is Downholland Moss, and it is therefore given the name DM-I. In Table 3, the mean altitude of the transgressive contact is -0.72 m O.D. derived from seven variates, and of the regressive contact -0.19 m O.D. derived from twelve variates.

The second transgression recorded in West Derby, DM-II, is clearly shown as a single lithologic unit in some of the borehole records on Downholland Moss, for example, DM-6, DM-7, DM-10, DM-15 and DM-16. Elsewhere on the Moss, a more complex sedimentary environment has been recorded during the period of the transgression and subsequent regression, as the records from DM-4, DM-5, DM-8 and DM-9 indicate. Two radiocarbon assays from Stratum 6 of DM-15 yielded dates that were too young for the beginning of DM-II, but the culminating stage of DM-II has been established at DM-15 by an assay on *Phragmites* peat from the base of Stratum 15, which yielded a date of 6050±65 radiocarbon years ago. Approaching marine conditions are indicated at DM-7 by low frequencies of Chenopodiaceae, *Plantago maritima* and *Artemisia*, following the regional elevation of the freshwater table and the encouragement of a rich reedswamp community, with Gramineae, *Typha angustifolia*, *Nymphaea*, *Galium*-type, *Lythrum salicaria* and *Filipendula*

TRANSGRESSION	TIME LIMITS YEARS B.P.	BOUNDARY	SITE. HEIGHT (METRES. ORDNANCE DATUM)	MEAN HEIGHT ±1σ
V	~2335	Sand dune palaeosol	FF-2, +5.08	+5.08
IV	(4800)-4545	Regressive	AM,+3.11; RH,+3.82; LL,+3.13; ATI,+3.40	+3.37±0.29
		Transgressive	AM,+2.90	+2.90
DM-III	(5900)-5615	Regressive	DM-2,+2.30; DM-3,+1.97; DM-4,+1.96; DM-5,+2.32; DM-6,+2.20; DM-7,+1.47; DM-8,+1.54; DM-9,+1.68; DM-10,+1.02; DM-16,+1.86; DM-16,+1.33; C-1,+1.96	+1.80±0.38
		Transgressive	DM-2,+0.87; DM-3,+1.23; DM-4,+1.21; DM-5,+0.99; DM-6,+0.65; DM-7,+0.93; DM-8,+0.32; DM-9,+0.50; DM-10,+0.60; DM-15,+1.27; DM-16,+1.17; C-1,+0.63.	+0.86±0.31
DM-II	(6500)-6050	Regressive	DM-2,+0.58; DM-3,+0.88; DM-4,+0.99; DM-5,+0.82; DM-6,+0.44; DM-15,+1.06; CM-1,+1.01; CM-3,+1.67; DM-4,+1.80; C-1,+0.52; CM-5,+1.86; CM-6,+1.54; CM-10,+1.44; CM-11,+1.20; C-1,+0.23	+1.07±0.50
		Transgressive	DM-4,+0.67; DM-5,+0.32; DM-6,+0.25; C-1,+0.14; CM-1,+0.27;	+0.33±0.18
DM-I	6980-6755	Regressive	DM-5,-0.26; DM-6,-0.04; DM-7,-0.02; DM-8,-0.38; DM-9,-0.61; DM-10,-0.36; DM-11,-0.40; DM-11A,-0.16; DM-15,-0.04; DM-16,+0.05; HH-6,-0.57; G-1,+0.37	-0.19±0.28
		Transgressive	DM-11,-0.87; DM-11A,-0.36; CM-1,-0.34; CM-10,-0.66; CM-11,-0.10; HH-6,-1.12; G-1,-1.57	-0.72±0.47
		Regressive	No data	
		Transgressive	LL,-10.21	-10.21

SITE CODES: FF - Formby Foreshore; AM - Alt Mouth; RH - Ranslett House, Formby; LL - Long Lane, Formby; G - The Gravel; ATI - Ash Tree Inn, Birkdale; DM - Downholland Moss; HH - Hillhouse; CM - Churchtown Moss; C - Crossens.

Table 3. Heights of Marine Transgressions in Lancashire: West Derby

represented. The succession is similar at DM-15, further seaward, and the beginning of marine conditions is anticipated within LPAZ DM-15a.

DM-II has been recorded on Churchtown Moss and at Crossens further north, and has mean values for the transgressive contact of +0.33 m O.D. derived from five variates and for the regressive contact +1.07 m O.D. derived from fifteen variates (Table 3). At DM-15, DM-II transgression is represented by a 76 cm thick unit (Strata 7 to 14) of blue silty clay, enriched with *Phragmites* rhizomes as the boundaries of the stratum are approached. There are significant changes in the particle size distribution through the stratum: coarse silts dominate the frequency distribution at about 170 cm, and above and below this level there are increases in the frequencies of medium and fine silts, and of the clay fraction which is dominant towards the boundaries of the stratum. Between 165 and 140 cm at DM-15, there is a significant decline in the frequency of both marine and marine-brackish diatoms, such as *Melosira sulcata* (Ehr.) Kützing, and an increase in the frequency of fresh-brackish water forms, such as *Diploneis ovalis* (Hilse) Cleve. Frequencies of both magnesium and calcium vary through the stratum, indicating changes of water depth and water quality (Tooley in the press).

The third marine transgression DM-III, is clearly shown on Fig. 33 as a stratum of silty clay, with occasional sandy and ferruginous partings. DM-III pushed landward as far as sampling site DM-10. The beginning of marine conditions is dated at DM-15 by an assay on *Phragmites* peat from the top of Stratum 15 of 5565±205 BP but may be too young and in error by up to 400 years. On the basis of the LPAZ DM-15d and DM-15e, the transgression ended some time before the Flandrian II/III chronozone boundary and an assay from DM-16 yielded a date of 5615±45 (Table 1, No 42) for the end of marine conditions.

At DM-15, the stratum of marine clay (Stratum 18) is dominated by fine silt and clay at the beginning and end of marine conditions, and during full marine conditions, between 75 and 85 cm in Stratum 18, there is an increase in the percentage frequency of coarse size fractions together with peak frequencies of calcium and magnesium (Tooley in the press). The diatom diagram from DM-15 (see Appendix) also shows a peak frequency of marine and marine-brackish diatoms at 85 cm, particularly represented by *Melosira sulcata* (Ehr.) Kützing, *Podosira stelliger* (Bail.) Mann and *Diploneis didyma* Ehrenberg. Approaching marine conditions, associated with DM-III, are shown in LPAZ, DM-7d, DM-lle and DM-15c. At DM-7, a reedswamp community with *Typha angustifolia*, *Galium*-type, *Lythrum salicaria*, *Filipendula*, *Hydrocotyle* and *Ranunculus*, and with open freshwater pools supporting floating aquatics, such as *Nymphaea* and *Potamogeton*, is replaced by salt marsh communities, characterised by Chenopodiaceae. At DM-15, there is a similar succession, with an enriched reedswamp community, less open freshwater and an increased frequency of salt marsh taxa such as *Plantago maritima* and *Artemisia*, probably explicable in terms of proximity to the coast. At DM-11, DM-III transgression is recorded indirectly by the regional elevation of the freshwater table within the perimarine zone. Strata 6, 7 and 8 comprise gyttja, and shallow, freshwater conditions are indicated by low frequencies of *Nymphaea, Nuphar* and *Myriophyllum* pollen, together with *Menyanthes* fruits. Fringing reedswamps of *Cladium mariscus, Typha angustifolia* and *Phragmites* are succeeded by an *Alnus-Salix* carr.

The end of marine conditions is shown most clearly at DM-16, where DM-III is recorded by a thin wedge of marine clay 16 cm thick recorded as Stratum 5. Minerogenic sedimentation in Stratum 5 is replaced by an 87 cm thick stratum of gyttja with *Menyanthes* fruits and woody detritus of *Alnus*. The progressive removal of marine conditions in this stratum is shown by the replacement within LPAZ DM-16a of salt marsh taxa, represented by the Chenopodiaceae and *Artemisia*, with freshwater and reedswamp taxa represented by *Typha angustifolia, Cladium mariscus, Lemna, Hydrocotyle, Iris* and *Lythrum salicaria*. A radiocarbon assay (Table 1, No. 42) yielded a date of 5615±45 BP which is older than the date for the commencement of marine conditions. However, it is thought that the earlier date is too young by about 400 years.

The transgression, DM-III, is also recorded at Crossens, and there are mean values of +0.86 m O.D. (Table 3) on the transgressive contact derived from twelve variates and +1.80 m O.D. on the regressive contact, also derived from twelve variates. It is significant that at DM-11, a freshwater gyttja was accumulating between +0.84 and +1.39 m O.D., *Menyanthes* fruits were recorded in this stratum between +1.12 and +1.07 m O.D., and in the pollen record there was a continuous *Nymphaea* curve.

Two additional marine transgressions can be inferred from stratigraphic, pollen analytical and radiometric evidence in the coastal area of West Derby. Transgression IV is recorded at the Alt Mouth, Ranslett House, Long Lane, Formby and at Ash Tree Inn, Birkdale. The transgressive boundary occurs at a height of +2.90 m O.D. and the regressive boundary at +3.37 m O.D. The justification for showing a distinctive transgression on Figs. 33 and 35 is that the altitudinal difference between DM-III and Transgression IV exceeds 1 metre (Table 3), and there is a lack of concurrence between LPAZs, ALMa and DM-15d (Fig. 4). Furthermore, biogenic sediments accumulating after Transgression IV at the Alt Mouth, are unequivocally post-elm decline, whereas those at DM-15 contain pollen assemblages that are both pre- and post-elm decline.

The transgression facies of Transgression IV comprise sands, silts and clays. At the Alt Mouth, there are strata rich in bivalves of *Cerastoderma edule* Linné, *Macoma balthica* (Linné) and *Barnea candida* (Linné), and on Formby Foreshore, the transgression is represented by sands and silts with *Scrobicularia plana* da Costa, *Barnea candida* (Linné), *Cerastoderma edule* Linné, *Astarte sulcata* (da Costa) and *Donax vittatus* (da Costa). (See also, Tooley 1970, 1977c).

A final transgression affecting West Derby can be inferred from the evidence from Formby Foreshore-3, where a date of 2335±120 BP was derived from an assay on woody detrital peat from a fossil dune slack at an altitude of +5.08 metres O.D. (Table 1, No. 67). Jelgersma *et al.* (1970) have argued that biogenic sedimentation in coastal dune areas in the Netherlands is closely related to periods of marine transgression, and dune instability and minerogenic sedimentation with periods of marine regression. If this argument applies equally to the Lancashire coast, then such an interpretation can be placed on the data from Formby Foreshore. Further support is lent to this argument when the age of the coastal dunes in West Derby is considered. On Downholland Moss, the seaward margin of the moss has been overblown by sand, and an assay on woody detrital peat with twigs of *Alnus* and *Betula* immediately below the over-burden of sand at DM-3 (see Fig. 6) yielded a date of 4090±175 BP; this coincided with a period of marine regression on the Lancashire coast (see Chapter 4, p. 136) post-dating Transgression IV.

The time limits and basis for separation of the four main transgressions recognised in West Derby are shown schematically in Fig. 35. They can be summarised as follows:

Transgression	Time Limits (Radiocarbon years, BP)
DM-I	6980 –6755
DM-II	(6500)–6050
DM-III	(5900)–5615
IV	(4800)–4545
V	–2335

Bracketed dates indicate that an age estimate has been made, based on stratigraphic, pollen analytic and altitudinal data.

LANCASHIRE: FYLDE IN AMOUNDERNESS

In the South Fylde, there is direct evidence of nine marine transgressions between 8570 and 1370 radiocarbon years ago, and indirect evidence of a tenth transgression from the sand-dune area of the coast. The evidence for these transgressions does not derive from a single site, but from four sites—Nancy's Bay, the Starr Hills, Lytham Common and Lytham Hall Park—all of which lie within the former township of Lytham. The marine sequences bearing this name serve as the type succession for Flandrian marine sequences in north-west England and the basis for inter-regional correlation (Figs. 35 and 39). A summary of the stratigraphic evidence, upon which the transgression sequences at Lytham are based, is given in Fig. 34, and detailed evidence from selected sites is given in Chapter 2.

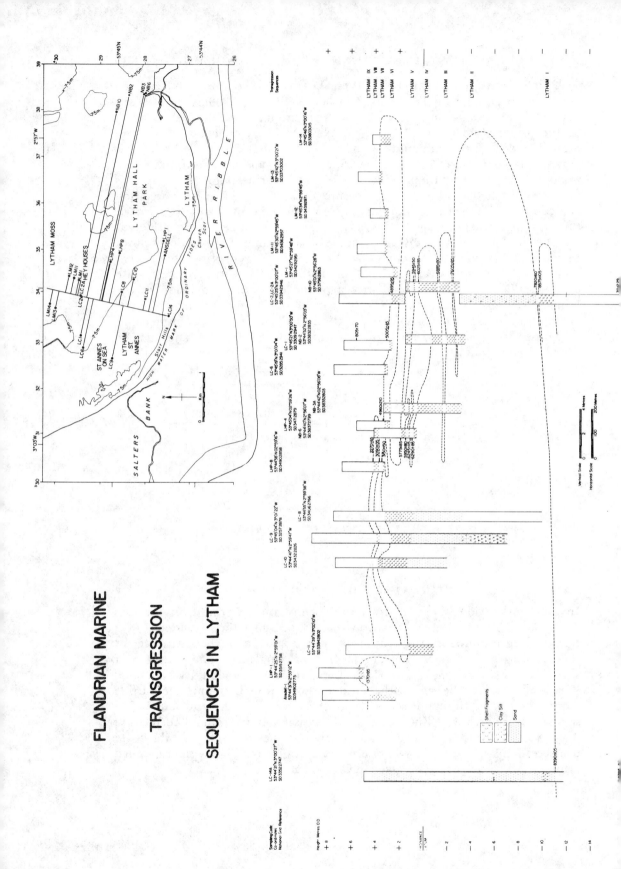

FLANDRIAN MARINE

TRANSGRESSION

SEQUENCES IN LYTHAM

The culminating stage of the first transgression to affect the Fylde, Lytham I, is recorded at LC-2A, Heyhouses Lane, St. Annes. Stratum 3 is a fine, grey clay layer with a few sandy partings and passes up into a black gyttja with occasional monocotyledonous remains. Within the gyttja, are low frequencies of open habitat, coastal taxa, such as Chenopodiaceae, *Plantago maritima* and *Artemisia*, but the overwhelming freshwater character of the sedimentary environment is shown by high frequencies of aquatic taxa, including *Typha latifolia, Myriophyllum alterniflorum, M. spicatum, Lemna* and *Hydrocotyle*. The sedimentation of marine facies ended at the site at an altitude of -9.75 m O.D. at or slightly before 8575±105 BP (Table 1, No. 58).

The second transgression to affect the Fylde, Lytham II, is recorded as a fine grey clay, and first crossed the line of the present coast at an altitude of -11.13 m O.D., at the Starr Hills, shortly after 8390±105 (Table 1, No. 55). Approaching marine conditions are anticipated in the woody detrital peat immediately subjacent to the grey clay by rising frequencies of aquatic taxa, particularly *Typha angustifolia*, and by a simultaneous increase in the frequency of open habitat, salt marsh taxa, such as the Chenopodiaceae and *Artemisia*. Lytham II penetrated further landward, and the transgression is recorded at LC-2A, Heyhouses Lane, at an altitude of -9.62 m O.D. a little after 7820±60 BP (Table 1, No. 57). Here also, approaching marine conditions are anticipated by a rise in the freshwater table, which encouraged the spread of *Typha angustifolia* and some Gramineae and brought about changes in the sedimentary régime from monocotyledonous peat to a gyttja. In the gyttja, immediately below the clay, there are increased frequencies of open habitat, coastal taxa, such as Chenopodiaceae, *Artemisia* and *Plantago maritima*.

There is, as yet, no evidence of the culminating stages of the second transgression, Lytham II, from the sedimentary sequences on Lytham Moss and Lytham Common, but at the eastern margin of Lytham Common, in Nancy's Bay, there is evidence of five marine transgressions between 7800 and 5700 radiocarbon years ago.

Lytham II is undoubtedly represented by the lowest grey silty clay recorded in Nancy's Bay (Figs. 16, 17 and 34). The beginning of the transgression has not been proved here, for the silty, marine clays were bottomed at only three sites (WOM-1, NB-9.5, NB-9.3 on Figs. 16 and 17), where till was encountered. The culminating phase of Lytham II is represented by the slightly undulating upper surface of the transgression, which has a mean altitude of -2.58 m O.D. derived from eight variates (Table 4). From the base of Stratum 2 at NB-10, in which high frequencies of open habitat taxa, such as Gramineae, Chenopodiaceae, Cruciferae, *Galium*-type, *Filipendula* and *Taraxacum*-type, have been recorded, an assay on the peat yielded a date of 6950±175 BP, which is not acceptable because of the concurrence of LPAZs NB-10a and RPAZ d at Red Moss; an estimated age of 7800 radiocarbon years is proposed.

Lytham III is recorded exclusively from Nancy's Bay. The transgressive

FIG. 34. Marine transgression sequences in Lancashire: Fylde in Amounderness: Lytham based on twenty-one cores from the south-west Fylde, in the vicinity of Lytham. The shading in each stratigraphic column is confined to those sediments of a marine and brackish-water origin. The pattern of marine transgression sequences established in north-west England is based on evidence from Lytham, and the scheme is shown here as Lytham I to IX. Differences in height of the transgression surfaces derive in part from the projection of each sampling site on to line A-A' shown on Fig. 25 and in part on a vertical exaggeration of x 50.

contact has a mean altitude of -2.51 m O.D. derived from eight variates and the regressive contact a mean altitude of -1.30 m O.D. derived from thirteen variates (Table 4). The transgression facies are a blue-grey silt, and there is a date for the beginning of the transgression at NB-10 of 7605±85 BP (Table 1, No. 48), which is supported by the concurrence of Local and Regional Pollen Assemblage Zones NB-10a and d at Red Moss. The culminating stages of Lytham III are recorded in Stratum 4 of NB-10, in which open habitat taxa dominate the pollen assemblage: Gramineae, Chenopodiaceae, *Artemisia*, *Galium*-type and *Filipendula* characterise the non-arboreal pollen assemblage, together with aquatic taxa such as *Lemna*, *Cladium mariscus*, *Typha angustifolia* and *T. latifolia*. An assay on *Phragmites* peat from the base of Stratum 4, yielded an age of 5880±180 BP (Table 1, No. 65) and is thought to be in error because of the consistent series of dates from higher levels in the core from NB-10 and from NB-6. A nominal age of 7200 radiocarbon years is given to the culminating stages of Lytham III.

Lytham IV has been established from stratigraphic evidence in the Lytham–Skippool valley, of which Nancy's Bay is at the southern limit. The mean altitude for the transgressive phase is -1.05 m O.D. derived from fourteen variates, and for the regressive phase +0.66 m O.D. derived from thirteen variates (Table 4). At NB-10, the transgression is heralded by a rise in frequency of open habitat taxa, especially the Chenopodiaceae, and occurs here at an altitude of -1.14 m O.D. some 6885±80 radiocarbon years ago (Table 1, No. 49). The transgression pushed northwards along the Lytham–Skippool valley, and is recorded at Peel-2, at an altitude of -0.80 m O.D. some 6535±110 radiocarbon years ago (Table 1, No. 44). At Peel, approaching marine conditions are indicated by an increase in the frequency of freshwater taxa, such as *Typha latifolia*, *T. angustifolia*, *Nymphaea* and *Hydrocotyle*, followed by a modest record of Chenopodiaceae.

The culminating stages of Lytham IV are shown by the complex sedimentary units recorded at NB-3A and -6, in Nancy's Bay. The stratigraphic successions show short-lived marine transgressions, separated by periods of brackish- and freshwater sedimentation. At NB-6, the culminating stages of Lytham IV comprise the following facies:

a. A coarse, sandy, grey-blue silt, passing up into a layer rich in *Phragmites* rhizomes.

b. An organic layer, 20 cm thick, with an open water facies at the base passing up to a terrestrial, monocotyledonous peat. Assays on thin slices of material from the bottom and top of this layer yielded ages of 6290±85 (Table 1, No. 54) and 6245±115 (Table 1, No. 53) radiocarbon years, respectively. *Typha angustifolia*, *Nuphar* and *Cladium mariscus* pollen were recorded in the gyttja, and Chenopodiaceae were recorded throughout the layer.

c. The organic layer passed up into a blue clay layer through a transition zone rich in *Phragmites*, and in turn to

TRANSGRESSION	TIME LIMITS YEARS B.P.	BOUNDARY	SITE. HEIGHT (METRES. ORDNANCE DATUM)	MEAN HEIGHT ± 1 σ
LYTHAM X	-817	Sand dune palaeosol / Regressive	LHP-6,+6.52; LC-1A,+5.63.	
LYTHAM IX	1795-1370	Transgressive	A-1,+4.02; A-2,+4.05; LHP-1,+5.39; LSV-1,+4.37	+4.46±0.55
		Regressive	No data	
LYTHAM VIII	3090-2270	Transgressive	LHP-5,+4.67; LHP-8,+4.19; NB-6,+4.15	+4.34±0.25
		Regressive	LHP-5,+3.67; LHP-8,+3.72; NB-6,+3.65	+3.68±0.03
LYTHAM VII	3700-3150	Transgressive	LHP-8,+3.51; LSV-3,+3.75; LSV-2,+3.65	+3.64±0.12
		Regressive	LSV-3,+3.53; LSV-2,+3.44	+3.49±0.05
LYTHAM VI	5570-4897	Transgressive	P-2,+2.34; LSV-2,+3.30; LSV-3,+3.28; LSV-4,+3.10; LSV-5,+2.42; LSV-6,+2.21; WM-1,+1.94; MyM-1,+2.14; MyM-2,+2.79; MyM-3,+2.28; LM-1,+2.69; LM-2,+3.30; LM-6,+3.46; LM-7,+3.54; LM-8,+3.65; LM-11,+3.27; LM-12,+3.30; LM-13,+3.70; LM-14,+3.68; LM-15,+3.26; LC-1,+2.94; LC-2A,+2.84; LC-2,+3.39; LC-6,+3.36; LHP-4,+3.21; LHP-5,+2.91; LHP-8,+3.49; MF,+2.95; L,+4.80.	+3.03±0.10
		Regressive	NB-3A,+2.84; NB-4,+2.15; NB-5,+3.65	+2.88±0.61
LYTHAM V	5947-5775	Transgressive	NB-12,+1.58; NB-11,+1.41; NB-10,+1.52; NB-9,+1.62; NB-8,+1.58; NB-7,+1.74; NB-1,+1.74; NB-2,+1.53; NB-6,+1.93	+1.59±0.10
		Regressive	NB-12,+1.38; NB-11,+1.22; NB-10,+1.22; NB-9,+1.23; NB-8,+1.34; NB-7,+1.24; NB-1,+0.97; NB-3A,+1.56; NB-6,+1.54	+1.30±0.17
LYTHAM IV	6710-6157	Transgressive	NB-12,+0.38; NB-11,+1.03; NB-10,+0.42; NB-9,+0.70; NB-8,+0.51; NB-7,+0.38; NB-6,+0.95; NB-2,+0.33; NB-3A,+0.59; NB-9.2,+0.86; NB-9.1,+0.92; NB-9.4,+0.71; NB-9.3,+0.79;	+0.66±0.23
		Regressive	NB-12,-1.11; NB-11,-0.94; NB-10,-1.14; NB-9,-0.71; NB-8,-1.12; NB-7,-1.26; NB-2,-1.55; NB-3A,-1.66; NB-1,-1.21; P-2,-0.80; NB-9.2,-0.94; NB-9.1,-0.94; NB-9.4,-1.07; NB-9.3,-0.50	-1.05±0.37
LYTHAM III	7605-7200	Regressive	NB-12,-1.22; NB-11,-1.05; NB-10,-1.27; NB-9,-0.77; NB-8,-1.22; NB-7,-1.38; NB-1,-1.26; NB-2,-1.75; NB-3A,-1.71; NB-4,-1.77; NB-9.1,-1.04; NB-9.2,-1.24; NB-9.4,-1.23;	-1.30±0.28
LYTHAM II	8390-7800	Regressive	NB-11,-2.44; NB-10,-2.40; NB-8,-2.61; NB-2,-2.66; NB-9.2,-2.63; NB-9.4,-2.04	-2.51±0.20
		Transgressive	NB-11,-2.50; NB-10,-2.48; NB-8,-2.71; NB-1,-2.75; NB-2,-2.74; NB-9.2,-2.71; NB-9.4,-2.08	-2.58±0.21
LYTHAM I	9270-8575	Regressive	LC-14A,-11.13; LC-2A,-9.62	-10.37±2.77
		Transgressive	LC-2A,-9.75	-9.75
		Transgressive	LC-2A,-9.97	-9.97

SITE CODES: A - Ansdell   LHP - Lytham Hall Park   LSV - Lytham - Skippool Valley   NB - Nancy's Bay   LC - Lytham Common   LM - Lytham Moss   P - Peel   MyM - Mythop Moss.   MF - Moss Farm   L - Lousanna

Table 4. Heights of Marine Transgressions in Lancashire: Fylde in Amounderness

d. A second organic layer, with a freshwater limnic deposit at the base and a terrestrial peat towards the top of the layer. An assay on material from the limnic deposit yielded a date of 6250±55 BP (Table 1, No. 69), and from the terrestrial peat a date of 5950±85 BP (Table 1, No. 52). This second organic layer is 21 cm thick, and freshwater reedswamp conditions are indicated by the pollen frequencies of *Typha angustifolia*, *Lemna*, *Peplis*, *Lythrum salicaria*, and *Galium*-type and open habitat, coastal taxa by Chenopodiaceae and *Plantago maritima*.

The effective end of Lytham IV is taken as the boundary between Strata 6 and 7 at NB-10, and between Strata 5 and 6 at NB-6.

Shortly after the end of Lytham IV, and more or less simultaneously throughout Nancy's Bay, there appears to have been a further transgression—Lytham V—which occurred at NB-6 at 5950±85 BP (Table 1, No. 52), and further landward at NB-10 at 5945±50 radiocarbon years ago, and ended at NB-6 at 5775±85 BP (Table 1, No. 47). The transgression occurred at a mean altitude of +1.30 m O.D. on the transgressive surface derived from nine variates and at +1.59 m O.D. on the regressive surface also derived from nine variates (Table 4), and at NB-6 comprised a 38 cm thick layer (Stratum 8) of blue clay, with black organic partings.

From Nancy's Bay, there is stratigraphic evidence of a final marine transgression—Lytham VI—for which there is a mean altitude value of +2.88 m O.D. derived from three variates (Table 4). The transgression facies in Nancy's Bay comprise a blue-grey clay, with rounded pebbles locally, some 30–80 cm thick. This layer also constituted the present cultivated surface of the Bay, and no radiometric assay could be attempted. The local pollen assemblage zones from the biogenic layers subjacent to Lytham VI are pre-elm decline, and, a late-Flandrian II age of 5570 radiocarbon years is proposed, and is supported by evidence from Downholland Moss in West Derby and Helsby Marsh in Cheshire.

There is abundant evidence of the culminating stage of Lytham VI, which occurred at a mean altitude of +3.03 m O.D. derived from twenty-nine variates (Table 4). There are two dates from Lytham Common of 5005±45 BP (Table 1, No. 40) and 4895±95 BP (Table 1, No. 56), one date from Lytham Hall Park of 4960±210 BP (Table 1, No. 23) and one date from the Lytham-Skippool valley at Peel of 4800±75 BP (Table 1, No. 43). From these dates and additional ones from Over-Wyre (Table 1, Nos. 25 and 59), a mean value for the end of Lytham VI of 4897 radiocarbon years is proposed. At LC-1, the end of Lytham VI is recorded by a blue-grey clay, rich in *Phragmites* rhizomes, passing up to a monocotyledonous peat with *Phragmites*. There is a progressive replacement of open habitat, coastal taxa, such as Chenopodiaceae, *Plantago maritima* and *Artemisia*, with freshwater taxa, such as *Typha angustifolia*, *T. latifolia*, *Hydrocotyle*, *Alisma* and *Potamogeton*. The removal of marine conditions, that records the end of Lytham VI, was widespread: the Lytham-Skippool valley was evacuated of water, and the western Fylde ceased being an island after a period of 1700 years: marine sedimentation ended throughout Lytham Moss and most of Lytham Common, and subsequent transgressions were recorded directly in Lytham Hall Park, a limited area to the south-west of the Park, and at the northern end of the Lytham-Skippool valley.

Lytham VII is represented in the Lytham-Skippool valley (see Fig. 21, LSV-2 and LSV-3) by a layer of tenacious, grey-blue clay with iron partings, comparable to a contemporary salt marsh soil. The layer occurred within 45 cm of the surface, and radiometric assays were not attempted. A date of 3700 radiocarbon years is proposed for the beginning of Lytham VII based on the stratigraphy, radiocarbon dates and altitude of dated material from Reed's Lane, Moreton, (see Table 1 No. 9 and Fig. 35). The end of Lytham VII is recorded in Lytham Hall Park, where at LHP-8 it is recorded at an altitude of +3.51 m O.D. at 3150±150 BP (Table 1, No. 22). At LHP-5 (Figs. 30 and 32), Lytham VII is represented by a 3 cm layer of clay (Stratum 5), with *Phragmites* rhizomes and other organic detritus throughout, and here, the transgression must be close to its marine limit. The pollen diagram from LHP-5 (Fig. 32) shows a recession of tree taxa at the level of the transgression, and an increase of herb and aquatic taxa. Of the herb taxa, Gramineae rise strongly, and there is an increase in the frequency of open habitat, coastal taxa, such as Chenopodiaceae, *Artemisia*, *Calystegia* cf. *soldanella*, *Armeria* and *Cynoglossum*. In addition, a single dinoflagellate cyst was recorded. Of the tree taxa, frequencies of *Quercus* increase immediately before the sedimentation of the clay layer, at the expense of *Alnus*. Freshwater taxa are represented by *Iris*, *Typha angustifolia* and *Hydrocotyle*.

The temporal separation of Lytham VII and VIII is very short. At LHP-8, they are separated by a layer of *Phragmites* peat 20 cm thick, and at LHP-5 this decreases to 9 cm (Stratum 6). The mean altitude of the transgressive phase of Lytham VIII is +3.68 m O.D., and of the regressive phase +4.34 m O.D. both derived from three variates (Table 4). The transgression facies at LHP-5 is a silty clay and at LHP-8 further seaward a silty sand. At LHP-4 80 m landward of LHP-5, the transgression is not recorded directly, but the deposition of a freshwater gyttja bears witness to the regional elevation of the freshwater table consequent upon Lytham VIII. At LHP-8, the beginning and end of Lytham VIII have been established by two radiocarbon assays of 3090±135 BP (Table 1, No. 21) and 2270±65 BP (Table 1, No. 20). The end of marine conditions appears to have been rapid, for open habitat coastal taxa, such as Chenopodiaceae and *Artemisia* are replaced by taxa closely associated with forest clearance, such as *Corylus*, *Salix*, *Plantago lanceolata*, *Taraxacum*-type and *Pteridium*.

In the south-west corner of Lytham Hall Park, and between the Park and the coast there is evidence of a ninth confined transgression. There are no data for the beginning of the transgression, but LHP-1 and Ansdell-1 and -2

furnish evidence of the culminating stages of Lytham IX. At LHP-1, a tenacious, grey-blue clay gives way to a woody detrital peat with *Phragmites* at
an altitude of + 5.39 m O.D. The pollen diagram (Fig. 31) gives unequivocal
evidence of the progressive removal of marine conditions. The lowest levels
analysed show high or persistent frequencies of Chenopodiaceae, *Plantago
maritima*, *Litorella*, *Artemisia* and *Ruppia*, and higher levels show a
preponderance of freshwater taxa, such as *Typha angustifolia*, *T. latifolia*,
Potamogeton and *Iris*. There was, unfortunately, inadequate material for a
radiocarbon assay, but the altitude of the lithologic boundary and the
distinctive pollen assemblage with Cereal pollen and weeds of cultivation
indicated a much more recent transgression than Lytham VIII. This was
confirmed when, at Ansdell -2, a radiocarbon assay on sandy, monocotyledonous peat yielded a date of 1795±240 BP (Table 1, No. 68). Pollen analyses
(see Fig. 29) from the assay level yielded an assemblage dominated by
herbaceous taxa of which open habitat, coastal taxa, such as Chenopodiaceae
and *Silene* were conspicuous. The removal of marine conditions permitted the
development of a hydrosere at the site, and frequencies of *Lemna*,
Hydrocotyle, *Potamogeton*, *Iris*, *Typha latifolia* and *T. angustifolia* are
recorded. At Ansdell -1, biogenic sedimentation following the end of marine
conditions was still under way at 1370±85 BP (Table 1, No. 66); these two
dates, together with their respective altitudes, establish the culminating stages
of Lytham IX.

A final transgression in the Fylde can be inferred from stratigraphic and
radiometric evidence from the coastal belt of sand-dunes. At Lytham
Common -1 and Lytham Hall Park -6, biogenic strata intercalating the blown
sand have been dated to 805±70 BP (Table 1, No. 61) and 830±50 BP (Table
1, No. 41), respectively. At LHP-6, pollen analyses yielded only *Taraxacum*-
type pollen, whereas at LC-1, there are low frequencies of *Alnus*, *Corylus*,
Quercus, *Populus* and *Pinus,* and high frequencies of dwarf shrubs, especially
Calluna, and of Cyperaceae. There is some evidence of an increase in wetness
prior to sand blowing, from the frequencies of *Hydrocotyle*, *Sphagnum* and
Salix. These data indicate a sand-dune slack origin for the biogenic stratum at
LC-1. If the argument of Jelgersma *et al.* (1970) is accepted, namely that
marine transgressions are closely associated in time with periods of dune
stability, increased precipitation and biogenic sedimentation in dune areas,
then the data from the coastal dunes of the Fylde coast indicate unequivocally
that a transgression was under way between 805 and 830 radiocarbon years
ago.

Documentary evidence from the Lytham Accounts in the muniments of the
Dean and Chapter of Durham Cathedral (Fishwick 1907; A. Piper pers.
comm.) suggests a period of relatively low sea-level from the mid-fifteenth
century until the early sixteenth century: the grazing land of Old Park,
Lytham, was overblown by sand in 1449-50, and the Green in 1464-5, 1476-8
until 1509-10. An alternative explanation could be found in the level and

extent of grazing on the coastal dune pastures during the latter half of the fifteenth century. But the persistence of unstable sand does indicate an extended intertidal zone from which the sand was recruited, consequent upon a relative fall of sea-level, and an increase in the frequency and strength of westerly winds.

In summary, the following ten marine transgressions are recognised in the Fylde of Lancashire. The time limits for some of the marine transgressions differ from those given elsewhere (Tooley 1974, 1976b, Huddart *et al.* 1977), because further radiocarbon dates have become available and permitted refinement of these limits.

Transgression	Time Limits (Radiocarbon years BP)
Lytham I	9270-8575
Lytham II	8390-7800
Lytham III	7605-7200
Lytham IV	6710-6157
Lytham V	5947-5775
Lytham VI	5570-4897
Lytham VII	3700-3150
Lytham VIII	3090-2270
Lytham IX	1795-1370
Lytham X	~817

CORRELATION OF COASTAL SEQUENCES IN NORTH-WEST ENGLAND

In north-west England, the most complete sequence of marine transgressions comes from Lytham (Fig. 35), which serves as the type area. Of the ten marine transgressions recognised at Lytham, homologues can be found elsewhere in Lancashire, in Cumbria and Cheshire. The Cumbrian evidence has been described elsewhere (Tooley 1969, 1974: Huddart and Tooley 1972, Huddart *et al.* 1977), and the evidence from Lancashire: Lonsdale and Cheshire in Tooley (1969, 1974). Table 5 and Fig. 35 show the correlation of marine transgressions in north-west England based on altitude, age and biostratigraphy.

There is at present no evidence for Lytham I in Lancashire: West Derby, but, on the basis of altitude, Lytham II has been identified at Formby. In West Derby, the beginning of marine sedimentation appears to have continued uninterrupted, at progressively higher altitudes, and to have penetrated far landward while two transgressions, Lytham III and IV, were under way in the Fylde. During Lytham IV, a single marine episode recorded in the Fylde, two marine transgressions were registered on Downholland Moss with altitudinal limits of -0.72 to -0.19 metres O.D. and +0.33 to +1.07

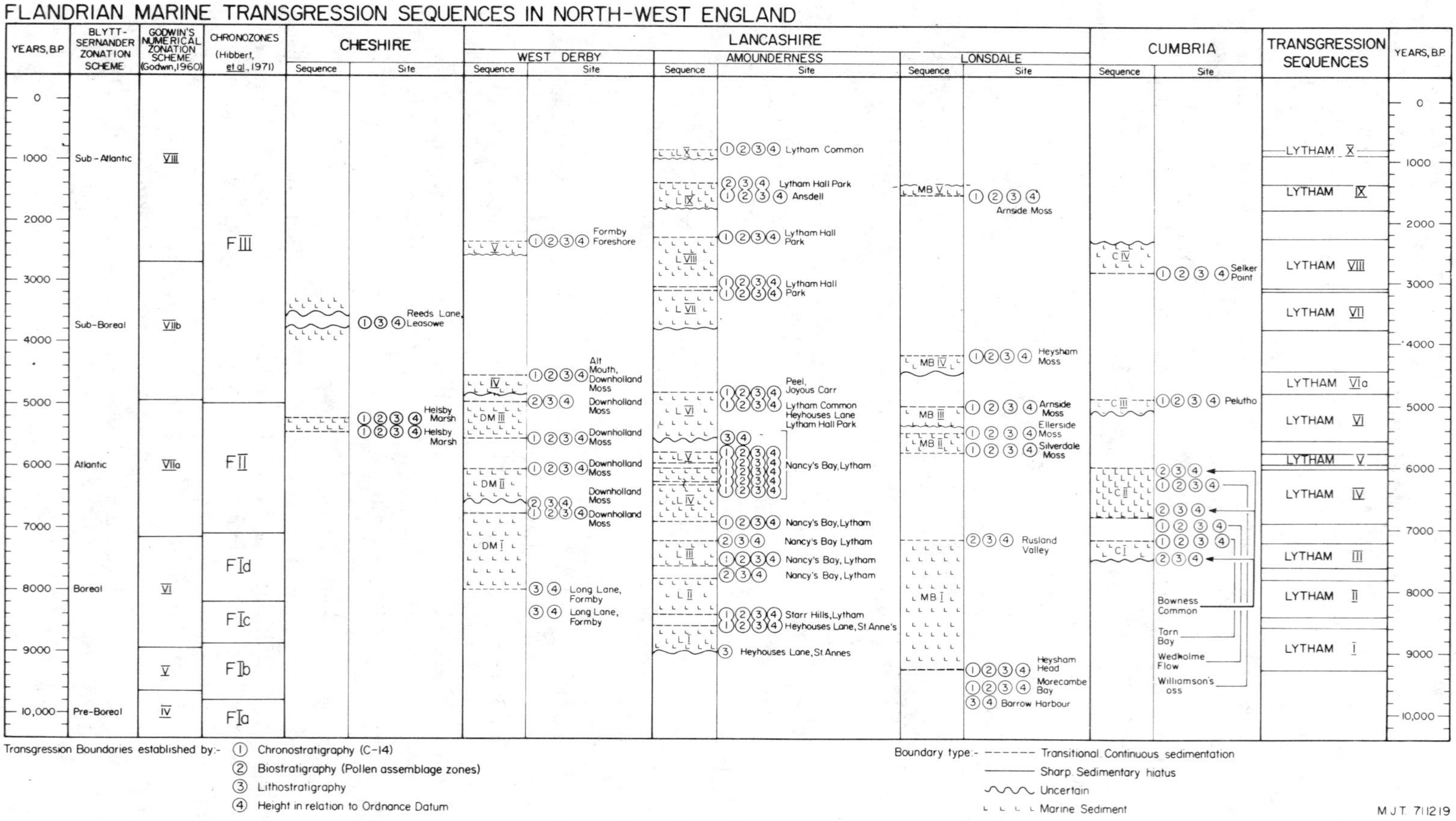

FIG. 35. Scheme of Flandrian Marine Transgression Sequences in north-west England. The temporal patterns of transgression sequences for Cheshire, Lancashire and Cumberland are shown, and, from the Amounderness area of Lancashire, the pattern of regional transgression sequences, LYTHAM I–IX is largely derived.

metres O.D. The altitudinal limits of Lytham IV are -1.05 to +0.66, which, taken with their standard errors, indicate that the same marine event is recorded. The two transgressions recorded in West Derby, Downholland Moss I and II, may register a local event, such as shoaling at the mouth of the Downholland Moss inlet, and the temporary exclusion of full marine conditions after 6750 years BP. However, the more enclosed and estuarine position of Nancy's Bay, from where Lytham IV is recorded, would be more susceptible to such a process, and none is recorded at this stage. An alternative explanation is that the radiocarbon dates establishing the time limits of Downholland Moss I have been rendered too young by post-depositional contamination, and that this transgression is homologous with Lytham III. This explanation is not, however, favoured, not only because of the concurrence of Local and Regional Pollen Assemblage Zones, but also because altitudinal data place both Downholland Moss I and II within the limits established for Lytham IV. In Table 5, these two transgressions are shown occurring whilst Lytham IV was under way in the Fylde. The culminating stage of Lytham IV is characterised in Nancy's Bay by a short-lived regression and transgression, such as those recorded in the stratigraphy of Downholland Moss-8, -9 and -10 (Figs. 6 and 33). Lytham V and Downholland Moss III are correlated on the basis of altitude and age.

Transgression	Years BP	Lancashire: Fylde Metres, O.D.	Lancashire: West Derby Metres, O.D.
Lytham X	~817	(+6.52) to (+5.63)	
Lytham IX	1795 - 1370	+4.46±0.55	
Lytham VIII	3090 - 2270	+3.68±0.03 to +4.35±0.25	(+5.08)
Lytham VII	3700 - 3150	+3.49±0.05 to +3.64±0.12	
Lytham VI	5570 - 4897	+2.88±0.61 to +3.03±0.51	+2.90 to +3.37±0.29
Lytham V	5947 - 5775	+1.30±0.17 to +1.59±0.10	+0.86±0.31 to +1.80±0.38
Lytham IV	6710 - 6157	-1.05±0.37 to +0.66±0.23	+0.33±0.18 to +1.07±0.50 and -0.72±0.47 to -0.19±0.28
Lytham III	7605 - 7200	-2.51±0.20 to -1.30±0.28	
Lytham II	8390 - 7800	-10.37±2.77 to -2.58±0.21	-10.21
Lytham I	9270 - 8575	-9.97 to -9.75	

Table 5. Correlation of marine transgressions in the Fylde and West Derby, Lancashire. Bracketted altitudes refer to the heights of palaeosols in coastal dunes.

Lytham VI constitutes a transgression that occurred during the culminating stages of Flandrian II and the beginning of Flandrian III. At some sites, for example, Lytham Common -1, the elm decline is underway as biogenic sedimentation replaced minerogenic sedimentation, whereas at others, for example Peel and Alt Mouth, biogenic sedimentation did not begin until after the elm decline. These sites either record a depositional hiatus, or register the passing of several centuries of continued marine presence before final evacuation of extensive tidal flat areas in West Lancashire by the sea. The fourth marine transgression in West Derby may be correlated with Lytham VI on the basis of both altitude and age.

A sand-dune palaeosol in Formby indicates that a marine transgression was under way, and although the altitude does not record the height of the marine event in West Derby, the date allows correlation with Lytham VIII.

In the Morecambe Bay area, Huddart *et al.* (1977) have summarised the evidence for five marine transgressions, and correlated them with the transgressions recorded at Lytham. The transgressive phase of Lytham I is recorded from Morecambe Bay, and the transgressive and regressive phases from the valley of the Rusland Pool, where Dickinson (1973) has described a marine clay with altitudinal limits of -0.75 to +1.35 m O.D. deposited during Flandrian Id.

On Silverdale, Helsington and Ellerside Mosses, a late Flandrian II transgression is recorded, and correlated with Lytham V.

Evidence for the Flandrian II/III transgression comes from Carnforth Levels, Arnside Moss and Helsington Moss.

The end of a fourth transgression is recorded on Heysham Moss, where both altitude and age individualise the transgression. Lytham VIa is identified here and in the Wirral, but, at present, there is no substantiating evidence from the type area. There is, however, some indirect evidence from Lytham that a transgression of limited extent and duration was under way. On Lytham Common (LC-1) and in Nancy's Bay (NB-6) there was a regional elevation of the water-table post-dating the elm decline, indicated by layers of *Menyanthes* fruits recorded in the stratigraphic descriptions and peak pollen frequencies of *Potamogeton, Typha angustifolia, T. latifolia* and occasional records of *Myriophyllum spicatum* and *Hydrocotyle.*

A fifth transgression, correlated with Lytham IX, is suggested by the evidence from Arnside Moss and Carnforth Levels.

In western Cumbria, the oldest marine deposit has been recorded from the Black Dub north of Allonby (Huddart *et al.* 1977), where a date on a sand-dune palaeosol indicates that during Flandrian Ic a marine transgression was under way. This event correlates with Lytham II.

Lytham IV is represented at Wedholme Flow, Bowness Common, Crosscanonby and Williamson's Moss, although marine sedimentation continued uninterrupted at Bowness Common and Wedholme Flow until Lytham V and VI respectively.

The Flandrian II/III transgression, Lytham VI, is represented from sites in the Duddon estuary. The raised shingle beach at Selker Point is associated with Lytham VIII.

ALTITUDINAL VARIATIONS OF MARINE TRANSGRESSION SURFACES

The differences in the recorded altitude of marine transgression surfaces in north-west England, Wales and south-west Scotland will give a measure of the amount of distortion of the surfaces since their formation attributable to differences in the amount of isostatic recovery.

In 1914, W. B. Wright showed the zero isobase of the so-called 25-foot raised beach following the southern coast of the Lleyn peninsula, touching the coast of the Wirral and bisecting the south-west Lancashire coast in the vicinity of the Alt Mouth. From these outer limits of the beach, the recorded altitudes of the raised beach rose northwards in the direction of maximum ice-loading, and therefore gave a measure of the magnitude of isostatic recovery. Gresswell (1953, 1957, 1958 and 1967) claimed to have mapped this raised beach throughout Lancashire and southern Cumbria, and named it the *Hillhouse coastline* (see Chapter 4). In Wales, Whittow (1960, 1965 and 1971) also recorded the raised beach, which he named the *Bryn-Carrog-coastline*. He argued for a revised position of Wright's zero isobase further south, in the vicinity of Borth, on the basis that at Porth Neigwl he claimed to have recorded a raised beach at a measured altitude of +3.6 m O.D. lying immediately north of the zero isobase. With slight modification, Churchill (1965b) has shown that biogenic deposits formed at sea-level 6500 radiocarbon years ago have been elevated subsequently from 3 to 9 metres above this level, and are now recorded at altitudes from 0.0 to +5.0 metres O.D. in Lancashire. Stephens and Synge (1966) have given graphical representation to these arguments, and show the limit of post-glacial raised shorelines close to the position of Wright's zero isobase. Furthermore, they suggest that in North Wales and Lancashire glacio-isostatic rebound is complete, and that, in consequence, recent post-glacial submergence, owing to the continued rise of sea-level, has taken place. Valentin's (1953) analysis of tide gauge data in the British Isles confirms this view.

In the light of these data, elevated beach deposits would be expected in North Wales, Cheshire, south-west Lancashire and the Fylde, and the recorded altitudes should increase from south to north.

In Tables 3 and 4 recorded altitudes of the transgression surfaces in West Derby and the Fylde are given. The point of measurement is the recorded altitude at which freshwater biogenic sedimentation gives way to brackish water or marine sedimentation in a marine transgressive sequence or vice versa in a regressive sequence. Table 5 summarises the mean altitudinal values of the transgression surfaces.

All the transgressions identified in the Fylde and West Derby can be isolated on the basis of altitude. Each transgression, except Lytham I, occurs at progressively higher altitudes in the stratigraphic sequence, and is individualised, not only by the altitudinal limits but also by mutually exclusive time limits. Furthermore, the altitudes of all the transgression surfaces occur below M.H.W.S.T., which intersects the coast at Formby at +4.15 m O.D. and at St. Annes-on-the-sea at +4.42 m O.D. The altitudinal data derive from a sample population of 197 variates, and the time data from 64 variates: greater confidence in the use of such data would derive from an increase in samples from both populations.

A comparison of the mean altitudinal values shows that the transgression surfaces are nearly horizontal and that no trend is discernible throughout the maximum latitudinal extent of the sites considered of 30 km.

Further south in Cheshire and North Wales, the near horizontality is maintained and extends the latitudinal extent of the area recording no significant gradients of the marine transgression surfaces to about 60 km. For example, Lytham VI is recorded at Rhyl in North Wales at an altitude of +2.43 m O.D. on the regressive surface (Table 1, No. 60) and well within a single standard error of the mean altitude for this transgression. The same transgression is recorded in Over-Wyre, north of the type area, at an altitude of +2.95 m O.D. (Table 1, No. 59), and at Peel, in the type area, at an altitude of +2.34 m O.D. (Table 1, No. 43).

The consistent altitudes of marine transgression surfaces recorded in the Fylde, West Derby, Cheshire and North Wales can be explained in the following terms:

a. Since the beginning of the first transgression, about 9300 radiocarbon years ago, the eustatic rise of sea-level has consistently exceeded any residual uplift caused by isostatic recovery. OR

b. The effects of the eustatic rise of sea-level have been amplified by down-warping along the eastern margins of the east Irish Sea Basin caused by hydro-isostasy. OR

c. Since Flandrian Ib, i.e. 9798 radiocarbon years ago, the amount of vertical movement has been negligible, and the sedimentary sequences in the tidal flat and lagoonal zones have registered eustatic changes directly.

The first explanation would result in thick deposits of medium to coarse sands, and there would be little opportunity for the development of alternating organic and inorganic facies: this pattern is found in Lonsdale and southern Cumbria. The second explanation is a viable alternative, but the evidence for downwarping such as the dipping of strata and minor dislocation with faulting has not been recorded; indeed, the evidence presented earlier indicates that over a latitudinal extent of 60 km across the East Irish Sea Basin the transgression surfaces maintain more or less uniform altitudes and are sub-parallel.

The possibility remains that, being marginal to the area of maximum ice loading, the Fylde, West Derby, Cheshire and north Wales suffered, firstly, elevation, and, secondly, depression, as areas further north became elevated isostatically, according to the theoretical model of Walcott (1972).

The evidence seems to indicate, however, that if these vertical movements occurred, they had ended by Flandrian Ib, which would support the final explanation.

The altitudinal data from the transgressions recorded in the Fylde and West Derby serve as a datum for comparing altitudes on transgression surfaces further north and south. Comparison, however, must be tentative because the provenance of the material, its altitude and its relationship to former positions of sea-level have not been established satisfactorily in adjacent areas, in Wales and Scotland.

LYTHAM I 9270–8575 years BP

The transgressive phase of Lytham I is registered at Port Talbot at 8970±160 (Q. 663) at an altitude of -19.03 metres O.D. The transgression may be recorded in the Trawling Grounds, Cardigan Bay, where peat dated to 8740±100 (Birm. 400) is overlaid at -18.5 m M.S.L. by marine sediments (Garrard 1977). In Morecambe Bay, the same transgression is recorded at Heysham Head at an altitude of -16.04 metres O.D. at 8925±200 (Table 1, No. 2), and at Irvine, Jardine (1964, and in Ergin et al. 1972) has recorded the same transgressive phase at an altitude of +6.2 metres O.D. at 8950±90 years BP (GU. 373). The gradient on the transgressive phase increases from 1 metre/100 km between South Wales and Morecambe Bay to 10.7 metres/100 km between Morecambe Bay and the Ayrshire coast. The transgressive phase of Lytham I has not yet been proved in the type area.

The regressive phase has so far not been established in Wales and the Somerset Levels nor in areas further north, but in the type area it is dated at 8575±115 and has an altitude on the isolation contact of -9.75 m (Table 1, No. 58).

LYTHAM II 8390–7800 years BP

The transgressive phase of Lytham II is probably registered at Dunball, Somerset, at 8360±145 (I. 4315) at an altitude of -20.7 metres O.D. and at 8360±140 (I.4403) at an altitude of -21.3 metres O.D. in the Somerset Levels (Kidson 1977). It is recorded in the type area at 8390±105 (Table 1, No. 55) where it has an altitude of -11.14 metres O.D. In Morecambe Bay, the same transgression is registered incontrovertibly at 8330±125 at an altitude of -16.03 metres O.D. (Table 1, No. 34). At Turnberry Bridge, Jardine (in Shotton and Williams 1971) has recorded the same phase at an altitude of +6.8 metres O.D. at 8420±150 (Birm. 190). During Lytham II, sea-level was rising very rapidly and part of the same transgression as a transgressive episode is registered towards the end of the period during which the transgression occurred. In the type area, this transgressive episode is recorded at 7820±60 (Table 1, No. 57) at an altitude of -9.65 metres O.D. In Morecambe Bay, it occurred at 7995±80 at an altitude of -11.16 metres O.D. (Table 1 No. 31), and on the north side of the Solway Firth, at Newbie Mains, Jardine (in Ergin et al. 1972) has recorded this phase of Lytham II at a height of +4.57 metres

O.D. for which there is a date of 7812±131 (GU. 375).

The transgressive phase of the earliest stage of Lytham II has a gradient of 3.3 m/100 km rising from Bridgwater Bay to Lytham and of 11.9 m/100 km *falling northwards* from Lytham to Morecambe Bay and a gradient of 11.7 m/100 km rising northwards from Morecambe Bay to the Ayrshire coast. This pattern is maintained towards the end of Lytham II, and a subsequent part of the same transgression has a gradient of 3.4 m/100 km, between Lytham and Morecambe Bay, *falling northwards* with less than half the earlier gradient, and a rising gradient of 16.7 m/100 km from Morecambe Bay to the north side of the Solway Firth, representing a considerable increase in gradient of the late Flandrian I transgression across the Lake District and into the Southern Uplands of Scotland.

LYTHAM III 7605-7200 years BP

The transgressive phase of this transgression has been established by three points: 7360±140 (I. 2690) at an altitude of -8.95 m O.D. in Bridgwater Bay (Kidson and Heyworth 1973), 7605±85 (Hv. 4125) at an altitude of -2.33 m O.D. in Nancy's Bay, Lytham (Table 1, No. 48) and 7540±150 (Birm. 222) at an altitude of +2.95 m O.D., at Newbie Cottages (Jardine, correspondence, and Shotton and Williams 1971). The gradient on the transgressive surface from Bridgwater Bay to Lytham is 2.7 m/100 km and this surface is subparallel to the transgressive surface of Lytham II. At Lytham there is a slight inflexion in the surface, and the gradient increases to the Solway Firth to 3.8 m/100 km. Unfortunately, there is no evidence for Lytham III at present in Morecambe Bay, and although there is a date of 7725±95 (Table 1, No. 29) on organic material immediately subjacent to marine clay at -14.84 metres O.D., Stratum 5 has been shown to constitute a reworked horizon. Furthermore, although the pollen assemblage corroborates the date on material from this stratum, it cannot be taken as an index of the relative position of sea-level in Morecambe Bay at the end of Flandrian I (See, Tooley 1974 and Huddart *et al.* 1977).

LYTHAM IV 6710-6157 years BP

Only the transgressive phase is available for comparison; the gradient between Bridgwater Bay and the type area is 1.7 m/100 km. This gradient is based on a comparison of an assay on a monocotyledonous peat, closely tied to a former position of sea-level in Bridgwater Bay (Kidson and Heyworth op. cit.) at an altitude of -5.95 m O.D., which yielded a date of 6890±120 (I. 2689), and an assay of monocotyledonous peat from Nancy's Bay (Table 1. No. 49) at an altitude of -1.17 m O.D. that yielded a date of 6885±80 (Hv.4126). It is probable that Lytham IV has been registered also at Newbie Cottages (Jardine 1971) on the north side of the Solway Firth, where peat subjacent to marine material has been dated at 7254±101 (GU. 64). Although the inauguration of the transgression here is older than further south, this is entirely consistent with the temporal pattern differences between isostatically unstable areas and marginal areas not suffering the same degree of recovery. The gradient between Nancy's Bay, Lytham and Newbie Cottages is 5 m/100 km and represents an increase in the amount of recovery in the Southern Uplands during the early part of Flandrian II. The regressive phase of Lytham IV is represented by silty clay from a borehole at Margam, South Wales, dated at 6189±143 (Q.275) at an altitude of -3.20 m O.D., by *Phragmites* peat from Downholland Moss (Table 1. No. 28) dated at 6050±65 (Hv. 3358) at an altitude of +1.06 m O.D., by *Phragmites* peat from Nancy's Bay (Table 1, No. 54) dated 6290±85 (Hv. 4131) at an altitude of +0.97 m O.D., and by a date on wood from the Palnure Borehole, Newton Stewart of 6240±240 (Birm. 189) at an altitude of +4.25 m O.D. (Jardine, correspondence, and Shotton and Williams 1971). Between West Derby and South Wales the gradient on the regressive surface is 1.8 m/100 km. There appears to be a slight dislocation of the surface between south-west Lancashire and the Fylde, but this is probably due to the open coast situation of the former, and the estuarine situation of the latter. Between the type area and the northern margin of the Solway Lowlands, the gradient on the regressive surface increases very slightly to 2.2 m/100 km.

LYTHAM V 5947-5775 years BP

The transgressive phase of Lytham V is represented by assays on woody detrital peat and *Phragmites* peat (Table 1, Nos. 16, 19 and 52). Here organic material immediately subjacent to marine sediments has been dated: the gradient on this transgressive surface is no more than 1.6 m/100 km, and is more or less parallel to the surfaces on Lytham IV in this area. Between Nos. 16 and 52 on Downholland Moss and in Nancy's Bay, the gradient is imperceptible. The regressive phase is well represented, and records a slight inflexion at Lytham, at the same point as Lytham III; further north, the gradient on the regressive surface increases. Between Bridgwater Bay and the type area, the regressive surface is registered by NPL.148, Q.382, Hv.3847 (Table 1, No. 42) and Hv.4128 (Table 1, No. 51). Q.382 is an assay on brushwood peat from Ynys-las, at an altitude of -0.46 m O. D., which yielded a date of 5898±135 (Godwin and Willis 1961). There is a gradient of 1.5 m/100 km on this surface; it increases to 4.2 m/100 km between the type area and the northern shore of the Solway Firth. This represents a slight reduction of gradient between Lytham III and Lytham V.

LYTHAM VI 5570-4897 years BP

The regressive contact of this Flandrian II/III transgression is recorded at Rhyl Beach (Table 1, No. 60), Heyhouses Lane (Table 1, No. 56) and Moss Farm (Table 1, No. 59). The surface shows a slight gradient within the type area and further south, a slope inflexion in the type area, and an increase in gradient northwards into the Solway Lowlands. Between the North Wales coast and the type area, the gradient on the regressive surface is 0.7 m/100 km; it increases to 4.8 m/100 km between the type area and the Solway Lowlands, indicating continued isostatic recovery over the Lake District and Southern Uplands.

LYTHAM VII 3700–3150 years BP

The regressive contact of Lytham VII is registered by Q.712 and Hv. 2918 (Table 1, No. 22), Q.712 is an assay on peat with roots of *Juncus maritimus* and *Phragmites* from Borth Bog at an altitude of +2.44 m O.D., which yielded a date of 2900±110 (Godwin and Willis 1962). A gradient on the surface of 0.6 m/100 km is recorded, and is sub-parallel to the regressive surface of Lytham VI.

LYTHAM VIII 3090–2270 years BP

Lytham VIII is registered at Lytham Hall Park-8 (Table 1, No. 20) where there is a date of 2270±65 (Hv.2916) on *Phragmites*-rich gyttja at an altitude of +4.22 m O.D. In Scotland, Jardine (1971) has recorded *Cardium* valves at an altitude of +5.24 m O.D., with an age of 2027±108 (GU.374, Ergin *et al.* 1972). The regressive surface of the transgression between these two index points maintains a minimum gradient of 0.7 m/100 km. If the same tidal relationship obtained between the type area and the Solway during the last 2000 years, then the Solway Lowlands have continued to rise isostatically, by an amount exceeding one metre.

LYTHAM IX 1795–1370 years BP

Lytham IX is based on a single dated stratum, and the regressive surface can be drawn parallel to Lytham VIII.

There is a considerable change in the altitude at which Mean High Water of Spring Tides intersects the coast of western Britain: at Burnham, the mean altitude is +6.2 m O.D., at Port Talbot +4.5 m O.D., at Aberdovey +2.3 m O.D., at St. Annes +4.4 m O.D., at Heysham +4.5 m O.D., and at Ayr +1.5 m O.D. (Admiralty Tide Tables 1970). The relevance of this surface resides in the fact that the dated regressive and transgressive contacts of the nine marine transgressions identified probably register a former position of M.H.W.S.T., and thus, in the area south of Morecambe Bay, all marine transgressions, except Lytham IX are now found below the contemporary M.H.W.S.T., whereas north of Heysham, the marine transgression surfaces rise progressively above the contemporary M.H.W.S.T. It appears therefore that the southern margin of Morecambe Bay is the area where the slope on marine transgression surfaces intersects the present effective spring tide level and that this relationship has been maintained throughout the Flandrian stage. South of Morecambe Bay, there is no evidence that the rise of sea-level has been overtaken by a rise in the land surface, and each slight oscillation in sea-level has been registered here. Whereas, further north, a more complicated relationship exists and sea-level has risen and fallen relative to the land surface, as continued land upheaval following deglaciation has been maintained at a decelerating rate.

The explanation of the behaviour of the transgressive surfaces of Lytham II in the vicinity of Morecambe Bay is elusive, and is probably to be found in the behaviour of the mantle in those areas marginal to maximum ice loading. Walcott (1972) has presented three models to explain the deformation of the earth, given the massive shifts of crustal load from the continents to the ocean basins and continental shelves during the Quaternary, involving periods of ice accumulation on land or rising sea-level, and concomitant delevelling. In the case of Morecambe Bay, it is possible to argue that it served as the southern limit for the culminating stages of the Devensian glaciation about 18,000 years BP, during which time the area further south was ice free. During the period up to 18,000 BP, both the Bride Hills moraine on the Isle of Man (Thomas 1977) and the moraine at Gutterby Spa (Huddart 1972, Huddart et al. 1977, Tooley 1977a) marked the effective southern limit of ice loading which, for the eastern shore of the Irish Sea, is much further north than Mitchell's estimate (1972). Movement of material within the asthenosphere southwards compensated for the loading north of Morecambe Bay, and brought about an elevation in the land surface in these recently deglaciated areas further south. Following deglaciation in the Lake District and Southern Uplands of Scotland, a return flow within the asthenosphere and transfer northwards to support the isostatic recovery there, would cause a subsidence in areas marginal to the area of maximum ice loading. Whilst these compensatory movements were under way, the eustatic rise of sea-level led to the inundation of Morecambe Bay during Flandrian Ib, and an additional surface load applied to an area recovering from the effects of ice loading. Walcott (1972) argues that one of the characteristics of hydro-isostatic displacement is a fall of land level seaward of the coastline, and it is possible that the application of an increasing water load in Morecambe Bay contributed to the distortation of the surfaces of Lytham II. It is perhaps significant that the marine transgressions in the Morecambe Bay area are not subdivided by biogenic sediments to the same extent as areas further south; this seems to indicate that, notwithstanding any uplift in this area, the eustatic rise of sea-level maintained this area as an embayment, and that only in small, enclosed areas, such as Arnside-Silverdale, did biogenic sedimentation intercalate inorganic marine sedimentation. Although the distance along a line orthogonal to the ice front across this area is probably too short, and the time for distortion also too short to register the transfer of mass within the asthenosphere and explain the distortion of Lytham II across Morecambe Bay, empirical studies by Krass and Ushakov (in Walcott 1972) have indicated uplift marginal to the limits of maximum crustal loading of 150–200 metres some 150 km outside these limits, and the effects significant but decreasing at distances up to 1500 km.

The distortion of Lytham II may reflect failure along the margin of the East

Irish Sea Basin, the limits of which are close to the sampling sites in
Morecambe Bay. Moseley (1972) has described post-Triassic earth movements
in north-west England and the re-activation of existing faults; he also shows two
faults crossing Morecambe Bay. The amplitude of the distortion is not greater
than 5 metres, and is maintained for a period not exceeding 500 years. Such a
short period certainly favours distortion caused by failure along the margin of
a sedimentary basin, rather than the more elaborate explanation involving
relaxation movements following deglaciation, although undoubtedly these
operated and require to be proved. If failure along fault planes occurred in
Morecambe Bay at some stage after Lytham II (and was probably completed
by Lytham III), then point loading of the crust from hydro-isostatic forces,
together with rapid sedimentation in the bay area following the rise of sea-
level, may have initiated the failure. There is no evidence elsewhere in north-
west England for such failure in the late Quaternary, and marine
transgression surfaces lack such distortion. However, in the Forth valley,
Sissons (1972) has recorded dislocation of raised shorelines, which he
interprets in terms of the re-activation of fault lines (see p. 23).

From the consideration of the behaviour of the marine transgression
surfaces in north-west England and adjacent areas, the following conclusions
can be made:

1. The surfaces are subparallel to parallel between the Fylde of Lancashire
and sites on the Welsh coast. Altitudinal differences are explained by the
point of measurement and tidal inequalities, and in most cases fall within the
standard error of the mean altitudinal value.

2. Each surface occupies its correct stratigraphic position between the Fylde
and Wales, and, except for Lytham IX, is below the contemporary Mean
High Water Mark of Spring Tides, at the altitude where it intersects the
present coast.

3. Increases in the gradients of the surfaces occur between the Fylde and the
southern margin of Morecambe Bay, and increase further between
Morecambe Bay and the Solway Lowlands.

4. The gradients of the transgression surfaces carry them above the
contemporary M.H.W.S.T. between Morecambe Bay and the Solway
Lowlands.

5. There is no evidence in the area north from Morecambe Bay to the Solway,
where elevated marine surfaces are recorded, that uplift has exceeded the
eustatic rise of sea-level until the middle of Flandrian III.

In the light of these conclusions, it is necessary to reconsider some of the
statements concerning isostatic recovery in north-west England.

The position of Wright's zero isobase, described earlier, cannot be substantiated. The present evidence indicates that the outer limit of raised shorelines
on the north-west coast of England is the southern margin of Morecambe Bay.
There is no unequivocal evidence of raised beaches of Late Devensian age
anywhere south of the Solway Firth, although Ashmead (1974) has assigned

some of the sea caves and benches around Morecambe Bay to the 'Paudorf Interstadial', which ended about 28,000 years BP (Tooley 1977b). Evidence for the position of sea-level prior to the opening of the Flandrian stage can be found in the Irish Sea Basins (Pantin 1977), the Isle of Man (Synge in the press; Thomas 1977) and areas further north that have suffered maximum ice loading. The southern limit of Wright's zero isobase should bisect the north-west coast of England 125 km further north in the vicinity of the southern shore of Morecambe Bay.

There does not appear to be any evidence to support Whittow's argument for shifting the isobase further south on the Welsh coast by about 150 km, and West (1972) has questioned the accuracy of the Porth Neigwl beach, upon which Whittow bases his argument. The field evidence for the *Hillhouse coastline* in Lancashire does not appear to exist (see Chapter 4), nor is the evidence for the *Bryn-Carrog coastline* in North Wales incontrovertible.

The limit for the post-glacial raised shoreline shown by Stephens and Synge for the north-west of England undoubtedly derives from Wright, Gresswell and Whittow, and cannot be substantiated in this position. Furthermore, their limit for the late glacial raised shoreline shown bisecting the Furness coast in south-west Cumbria has no validity; there do not appear to be any transgressions of greater antiquity than Flandrian Ib recorded along the North Lancashire and Cumbrian coasts, and this outer limit must be pushed northwards by up to 100 km, thereby supporting the conclusions of Sissons (1967). There is some evidence from the Isle of Man (Thomas 1977) and north-west England (Tooley 1977b) of glacio-marine sediments of Late Devensian age, and the possibility of a marine accompaniment to deglaciation with associated Late Devensian shorelines at $+14$ to $+18$ m O.D. needs further investigation.

Finally, there is little support for Churchill's argument (1965b): biogenic deposits that formed close to sea-level some 6500 radiocarbon years ago in south-west Lancashire are at present recorded at altitudes below zero O.D. and not from 0 to 5 metres above Ordnance Datum.

This description of the altitudinal relationships of the marine transgression surfaces along the western margin of Britain permits not only a reappraisal of the field evidence in the critical area of north-west England, but also the identification of areas where elucidation is needed. These include the coast from Heysham to the Solway Firth, and the whole of the north and west coasts of Wales, all of which contain sites worthy of further attention.

CHAPTER 4

Relative Changes of Sea-Level in North-West England

The geographical and temporal pattern of marine transgressions established in north-west England (see Chapter 3) is an expression of relative sea-level changes that have affected this part of the British coast.

Relative sea-level changes record not only global (eustatic) movements of sea-level, but also regional crustal movements attendant upon glaciation (glacio-isostasy), changes in water load on the continental shelf (hydro-isostasy) and subsidence within and adjacent to sedimentary basins. Sediment consolidation will also affect the altitude of deposits after burial. Both geographical and temporal scales are critical in making an assessment of the importance of each of these factors that affect the magnitude of relative sea-level changes. The assignment of actual quantities to each of the factors would allow the solution of a sea-level formula in which a measure of eustasy was sought. There is, however, no formula that integrates satisfactorily all the elements that contribute towards an explanation of the present-day altitude of fossil marine deposits. It is therefore more satisfactory to think in terms of relative sea-level changes, in which all the factors contribute, in varying proportions, on a finite time scale. An estimate can be made of the overriding element affecting the altitude of fossil marine deposits, but, it remains a qualitative estimation.

For north-west England three graphs have been drawn to indicate the course of relative sea-level changes during the Flandrian stage. Fig. 36 is a time-depth graph, upon which all the variates related directly or indirectly to sea-level changes in north-west England have been plotted: the standard error for each radiocarbon date is shown, and an altitudinal deviation is also given for those dates whose provenance within the sampled layer is not known precisely. Fig. 37 again shows each variate, but with an indication of the position of the dated sample in relation to a marine deposit: thus, an arrow pointing upwards indicates a dated level immediately below a marine deposit, and an arrow pointing downwards indicates a dated level immediately above a marine deposit — a convention adopted by Donner (1970) with effect. The former shows that shortly after the date, a marine transgression was under way; the latter gives evidence for a marine regression. The continuous line curve on Fig. 37 traces the movement of Mean High Water Mark of Spring Tides and the pecked line curve traces the movement of Mean Tide Level. The amplitudes and periods of the oscillations on the curve have been established by the twenty-six index points shown in Fig. 38 from a limited area in west Lancashire and considered in detail in the succeeding section and in summary elsewhere (Tooley 1976b). An estimate is made of the relationship of each

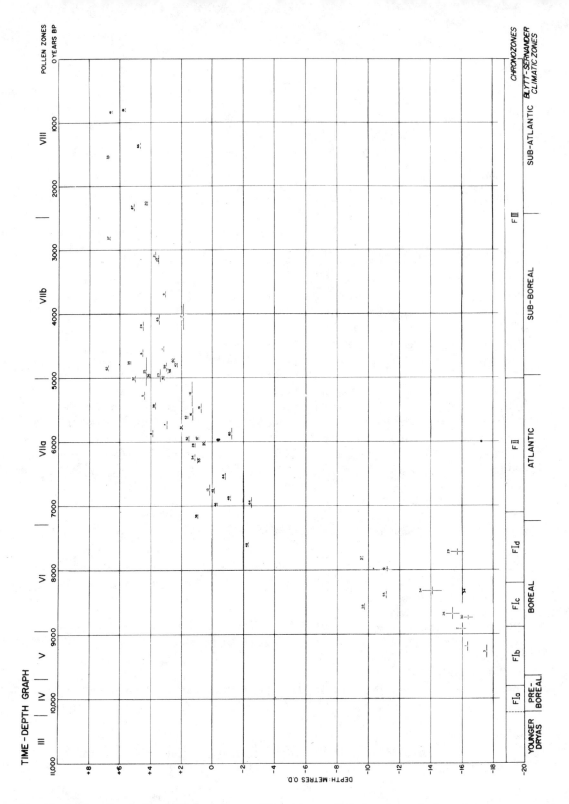

Fig. 36. A Time-Depth Graph from north-west England, on which dated samples have been plotted with the standard deviation of the date and the depth deviation. The numbers refer to the list of radiocarbon dates given in Table 1.

index point to a former position of sea-level, and of the change in the altitude of the sample caused by consolidation.

THE RELATIVE SEA-LEVEL CURVE (Figs. 37 and 38)

Of the twenty-six index points used to construct the sea-level curve, twenty-five derive from sites adjacent to the Mersey and Ribble estuaries, and one from Morecambe Bay. The twenty-five index points come from a restricted coastal area in west Lancashire, with a latitudinal extent no greater than 60 km and are adjacent to estuaries with similar tidal ranges at the present day. Furthermore, the differences in altitude of the marine transgression surfaces in west Lancashire are slight, and, throughout the Flandrian stage, range from 0.02 to 1.79 m (see Chapter 3). The use of a single index point from Morecambe Bay, some twenty-eight km north, does mean that this altitude is not strictly comparable with the altitudes of the other index points. But the justification for using it is twofold: the point records the lowest sea-level altitude in north-west England, and the present-day tidal range at Heysham is intermediate between the tidal ranges at St. Annes and Liverpool.

Point 1. Heysham Head M2.13. 8925+200 (Birm. 140) -16.04m. O.D. (Table 1.
No. 2).
The index point derives from an assay on hard, dry, brown, monocotyledonous
peat from the middle of the biogenic layer (Stratum 2) and there is an
altitudinal deviation of +0.65m. The beginning of marine conditions occurs
0.63m. above the dated level, which is therefore maximal. In the pollen
record there is some evidence of approaching marine conditions in the low
frequency of Chenopodiaceae. The biogenic layer was 1.60m. thick and was
underlaid by a stiff, reddish brown till: some consolidation of the biogenic
layer has undoubtedly taken place. The peat is overlaid by a blue-grey,
silty clay coarsening upwards to a buff, silty, fine sand with shell fragments.
Additional points, shown on Figure 39 from Morecambe Bay, are not used as
index points for the sea-level curve, because either the radiocarbon date was
not supported by pollen analytical evidence, or approaching marine conditions
were not recorded in the pollen record.

Point 2. Heyhouses Lane, St. Annes. LC-2A. 8575+105 (Hv. 4346) -9.75m. O.D.
(Table 1. No. 58)
A relative fall of sea-level is inferred from the data at Heyhouses Lane. The
assay was on gyttja from the base of an organic layer (Stratum 4) immediately
above a fine, grey marine clay, with sandy partings. The progressive removal
of marine conditions is indicated weakly by the replacement of open habitat,
coastal taxa, such as Chenopodiceae and *Artemisia*, by freshwater aquatic taxa,
such as *Typha latifolia*, *T. angustifolia*, *Myriophyllum spicatum*, *M. alterniflorum*,
Lemna and *Hydrocotyle*.

Point 3. The Starr Hills, Lytham. LC-14A. 8390+105 (Hv.4343) -11.14m. O.D.
(Table 1. No. 55).
A relative rise of sea-level is indicated from the Starr Hills, where biogenic
sedimentation is replaced by minerogenic sedimentation of marine origin at
-11.13m. O.D.   An assay on woody detrital peat with charcoal, immediately

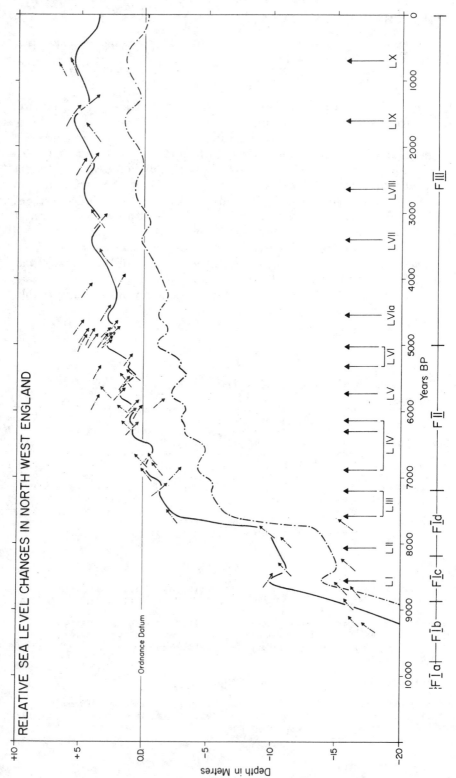

FIG. 37. A graph to show relative sea-level changes in north-west England. An arrow pointing upwards indicates a dated sample immediately below a marine deposit, and an arrow pointing downwards a dated sample immediately above a marine deposit. The continuous line curve shows the change in altitude of the spring tide level, whereas the pecked line curve shows the movements of mean tide level. L-I to L-X are marine transgressions recorded at Lytham.

below the marine clay, yielded a date that was corroborated by pollen analyses
(Fig. 26). Approaching marine conditions are indicated strongly at the dated
level by rising frequencies of Gramineae, Chenopodiaceae, and *Artemisia*.
The biogenic layer (Stratum 3) is 8 cm. thick and overlies, in turn an organic
sand and a hard, tenacious pink clay. Consolidation of the biogenic layer
would have been slight, and the recorded altitude of the biogenic/marine
clay boundary is probably close to its altitude at the time of burial during
Flandrian Ic.

Point 4. Heyhouses Lane, St. Annes. LC-2A. 7820+60 (Hv. 4345) -9.65m. O.D.
(Table 1. No. 57)
An assay on gyttja from the top of a biogenic deposit yielded a date that
records the beginning of marine sedimentation at -9.66m. O.D. The pollen
assemblage (Fig. 27) shows the replacement of aquatic taxa, such as *Typha
angustifolia*, by open habitat coastal taxa, such as *Plantago maritima*,
Chenopodiaceae, and *Artemisia*. The radiocarbon assay is corroborated by
the pollen assemblage. The recorded altitude of the assay level has been
lowered as the result of consolidation of Strata 4, 5 and 6.

Point 5. Nancy's Bay -10, Lytham. NB-10. 7605+85 (Hv. 4125) -2.33m. O.D.
(Table 1. No. 48).
The beginning of a marine transgression in Nancy's Bay is recorded by this
point, which is a date on monocotyledonous peat, with *Phragmites* rhizomes
at the base of a layer of marine silt (Stratum 3). The presence of open
habitat, coastal taxa, such as Chenopodiaceae and *Artemisia* testifies to
the proximity of marine conditions. The recorded altitude of the transition
from a freshwater to brackish water palaeoenvironment at -2.40m. O.D., has
been lowered by consolidation of the 8 cm. thick layer of monocotyledonous
peat, but the amount is probably slight, because of the narrowness of the
biogenic layer and the presence beneath it of a dense layer of coarse silt.

Point 6. Nancy's Bay -10, Lytham. NB-10. 7200 (estimated) -1.27m. O.D.
An assay on a 3 cm. thick sample of monocotyledonous peat with *Phragmites*
from the base of a 13 cm. thick biogenic layer (Stratum 4) overlying marine
silt, should have provided a date for the end of the transgression, and a
relative fall of sea-level. The assay yielded a date of 5880+180 (Hv. 4707)
which was unacceptable on the grounds that the empirical limits of *Alnus*
were recorded in Stratum 4, diagnosing the Flandrian I/II Chronozone boundary:
an estimated age of 7200 radiocarbon years has been assigned to Point 6.
The undoubted proximity of marine conditions is shown by the pollen assemblage
NB-10b (Fig. 19), and the recorded altitude of the end of marine sedimentation
at -1.27m. O.D. is little affected by consolidation: the stratum of marine
sediments is 113 cm. thick and comprises a medium to fine silt with coarse
silt and clay partings throughout. At Bootle Beach, Cumbria (Table 1.
No. 38), the end of the marine conditions is dated at 7160+75 (HV. 3843) and
even though with a recorded altitude of +0.91m. O.D. registering continued
uplift in this area, this date confirms the downward tendency of sea-level.

Point 7. Downholland Moss -11A DM-11A. 6980+55 (Hv. 3936). -0.38m. O.D.
(Table 1. No. 46).
An assay on a 3 cm. thick sample of sandy, monocotyledonous peat with
Phragmites yielded an early Flandrian II date, which was corroborated by
the pollen assemblage. The biogenic layer (Stratum 3), from which the
sample was taken, was only 3 cm. thick and overlaid a coarse sand. The
recorded altitude of the transition from biogenic sedimentation to the
overlying layer of marine clay, of -0.36m. O.D. must be close to the altitude
to which sea-level rose and inundated the site, for the consolidation rate
of an organic, coarse sand is approaching zero. Approaching marine
conditions are indicated by a short-lived rise of the pollen of aquatic taxa
(Fig. 8), such as *Sparganium, Typha angustifolia, Potamogeton*, and *Nymphaea*,
followed by an increase in the frequency of open habitat, coastal taxa,
such as Chenopodiaceae, *Armeria, Hippophae* and Tubuliflorae, such as *Artemisia*.

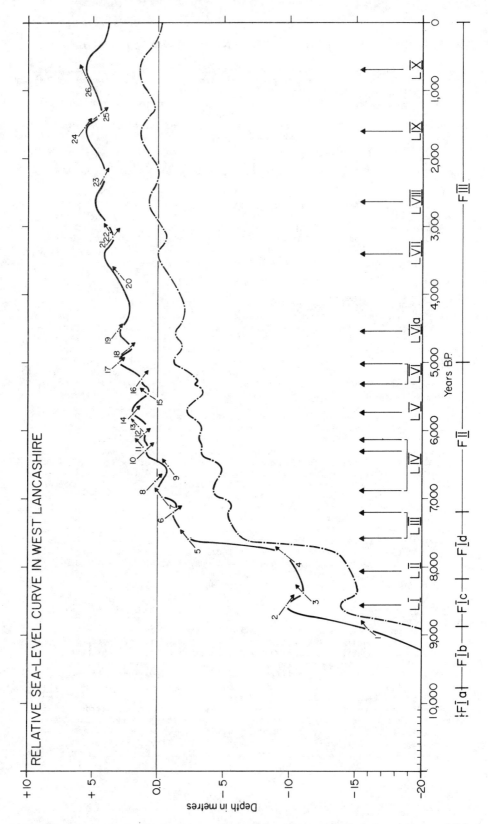

Fig. 38. Relative sea-level curve in west Lancashire. The twenty-six index points establish the amplitude and period of sea-level oscillations in a restricted area in west Lancashire.

The rise of sea-level recorded by Point 7 on Downholland Moss, is also recorded on Nancy's Bay, but at a lower altitude. Here, marine sedimentation began shortly after 6885+80 (Hv. 4126) at an altitude of -1.14m. O.D. (Table 1. No. 49).

Point 8. Downholland Moss -11A. Dm-11A. 6760+95 (Hv. 3935) -0.14m. O.D. (Table 1. No. 45).
A relative fall of sea-level is recorded by Point 8. A 4 cm. slice of gyttja with *Phragmites* from a biogenic layer, immediately above marine clay, yielded a date that was supported by the local pollen assemblage zone. Pollen analysis at the dated level (Fig. 8) shows the beginning of an autogenic succession from a salt marsh community with *Plantago maritima, Artemisia, Armeria* and Chenopodiaceae to a freshwater community with *Typha angustifolia, Cladium mariscus* and *Nymphaea*. The fall in sea-level is corroborated by a date of 6750+175 (Hv. 2680A) at an altitude of +0.15m. O.D. (Table 1. No. 13) on a brackish water clay gyttja immediately above a silty clay of marine origin. The recorded altitude of the isolation contact at DM-11A of -0.16m. O.D. and at DM-15 of +0.04m. O.D. has not been affected greatly by consolidation since burial.

Point 9. Peel -2A. P-2A. 6535+110 (Hv. 3934). -0.83m. O.D. (Table 1. No. 44).
A sample of woody detrital peat was taken from the top of a 27 cm. thick layer (Stratum 3) overlying till, and immediately subjacent to a 314 cm. thick layer of blue-grey, silty estuarine clay. The recorded altitude of the isolation contact was at -0.80m. O.D., and some consolidation of Stratum 3 has resulted in a lower recorded altitude. Approaching marine conditions in the Lytham-Skippool valley are shown weakly by low, persistent frequencies of open habitat taxa, such as Chenopodiaceae, *Rumex, Galium, Stellaria* and *Filipendula*.

Point 10. Nancy's Bay-6. NB-6. 6290+85 (Hv. 4131). +0.97m. O.D. (Table 1. No. 54).

Point 11. Nancy's Bay-6. NB-6. 6245+115 (Hv. 4130). +1.13m. O.D. (Table 1. No. 53)
Points 10 and 11 come from the bottom and top of a biogenic layer. (Strata 3 and 4) in which a freshwater gyttja gives way to a telmatic *Phragmites* peat. The strata are both underlaid and overlaid by clays of marine origin. Throughout the short period of biogenic sedimentation, the pollen diagram (Fig. 18) indicates the proximity of open habitat coastal taxa. Point 10 records a relative fall in sea-level and Point 11 a relative rise. 18 cm. above Point 11, an additional radiocarbon assay dates the end of the deposition of the marine clay (Stratum 5) at 6250+55 (Hv. 5294), and, there is a recorded altitude for the isolation contact of +1.35m. O.D.
These data indicate a short period during which relative sea-level rose to a maximum value of +1.35m. O.D., but the pattern of transgressions before this maximum was reached probably reflects local site conditions, such as sediment consolidation.

Point 12. Downholland Moss-15. DM-15. 6050+65 (Hv. 3358) +1.06m. O.D. (Table 1. No. 28)
Point 12 records a relative fall of sea-level and the end of the second transgression in west Derby, DM-II. The sample comprised a 3 cm. thick slice of *Phragmites* peat from the base of a biogenic layer (Stratum 5). The stratum was underlaid by 76 cm. of blue, silty clay of marine origin, and consolidation would have been minimal: the recorded altitude of the isolation contact at +1.06m. O.D. is thought to be close to the altitude at which marine conditions ended on Downholland Moss at the middle of Flandrian II. A progressive vegetation succession is indicated from the pollen diagram (Fig. 9), in which open habitat, coastal taxa, such as Chenopodiaceae, and *Artemisia,* are replaced by freshwater, aquatic taxa, including *Typha angustifolia*. In Nancy's Bay, Lytham, a fall of sea-level

is also recorded at this time: at NB-10, an assay on a *Phragmites*-rich clay,
at the top of a marine clay layer (Stratum 6) yielded a date of 6025+85
(Hv. 4127) at a recorded altitude of +0.45m. O.D. (Table 1. No. 50). A
progressive vegetation succession is strongly developed (Fig. 19):
herbaceous taxa, such as Gramineae, Cyperaceae, Chenopodiaceae and *Artemisia,*
are replaced by freshwater taxa, such as *Typha angustifolia, T. latifolia,*
Lemna, Cladium, Myriophyllum spicatum, Hydrocotyle, Potamogeton and *Iris.*

Point 13. Nancy's Bay-6, Lytham. NB-6. 5950+85 (Hv. 4129) +1.52m. O.D.
(Table 1. No. 52)
The beginning of a short-lived transgression, and a relative rise of sea-level
is recorded by Point 13 from Nancy's Bay. The assay was on 4 cm. of
Phragmites peat from the top of a 15 cm. thick biogenic layer (Stratum 7).
The recorded altitude of the isolation contact at +1.54m. O.D. has been affected
by consolidation of the *Phragmites* peat. Approaching marine conditions
are indicated by the pollen analysis from the biogenic layer (Fig. 18):
there is a reduction of arboreal pollen and a rise of herbaceous pollen
contributed by Gramineae and Cyperaceae. Freshwater taxa, such as *Typha*
angustifolia, Potamogeton, and *Lemna,* are superseded by open habitat, coastal
taxa, such as *Plantago maritima, Artemisia* and Chenopodiaceae.
At Nancy's Bay-10, landward of NB-6, the rise of sea-level is recorded by an
11 cm. layer (stratum 9) of brackish-water clay, rich in *Phragmites*. A
radiocarbon assay from the top of this layer (Table 1. No. 47) yielded a date
of 5945+50 (Hv. 4124). The recorded altitude of the rise of sea-level at
NB-10 is +0.76m. O.D., and, as the marine deposit overlies a 31 cm. thick
layer of partly humified, monocotyledonous peat, with *Phragmites,* it is
probable that consolidation has carried down the altitude of the isolation
contact by at least 78 cm. compared with NB-6.

Point 14. Nancy's Bay-6, Lytham. NB-6. 5775+85 (Hv. 4128). +1.93m. O.D.
(Table 1. No. 51).
A relative fall of sea-level is recorded by Point 14, which is a date on a
4 cm. sample of *Phragmites* peat, immediately above marine clay. The recorded
height of the isolation contact is at +1.92m. O.D. Pollen analysis from the
base of the biogenic layer (Stratum 9) showed a progressive vegetational
development from open habitat, coastal taxa, with Chenopodiaceae, and *Artemisia,*
to freshwater aquatic taxa including *Typha angustifolia, Lemna, Cladium* and
Hydrocotyle, to a woody fen with *Quercus, Alnus, Salix, Euonymus* and *Frangula*
and a herb layer of *Filipendula, Lythrum salicaria, Galium, Osmunda* and
Polypodium. The recorded altitude of the isolation contact, overlying marine
clay, has probably not been affected greatly by consolidation. Corroborating
evidence for a fall in sea-level at this time comes from Silverdale Moss and
Downholland Moss. At Silverdale Moss, marine sedimentation ended at 5865+115
(Q.216) at an altitude of +3.85m. O.D., (Table 1. No. 8) which is a measure of
the amount of isostatic recovery in this area since late Flandrian II and the
fall in sea-level has undoubtedly been advanced temporally for this reason.
On Downholland Moss, a fall of sea-level is recorded shortly before 5615+45
(Hv. 3847) which is a date on gyttja 14 cm. above the isolation contact (Table
1. No. 42). The recorded altitude of the isolation contact here is +1.33m.
O.D. and has been affected by consolidation to a certain extent: the gyttja
overlies, in turn, a marine clay layer 16 cm. thick and a layer of *Phragmites*
peat, now 15 cm. thick. The pollen diagram from DM-16(Fig. 10) shows the
replacement of open habitat, coastal taxa, such as Chenopodiaceae and *Artemisia,*
by freshwater aquatic taxa, such as *Cladium, Typha angustifolia, Lemna,*
Hydrocotyle and *Iris.*

Point 15. Helsby Marsh-1. HM-1. 5470+155 (Hv. 2686). +0.73m. O.D.
(Table 1. No. 19).
Point 16. Helsby Marsh-1. HM-1. 5250+385 (Hv. 2685). +1.29m. O.D.
(Table 1. No. 19).
Points 15 and 16 are taken together because they record a relative rise and
fall of sea-level on Helsby Marsh, south of the Mersey estuary. Point 15

is an assay on woody detrital peat from a 10 cm. layer of peat overlying in turn coarse sand and till: the recorded altitude of the beginning of marine sedimentation at +0.73m. O.D. has not been affected greatly by consolidation. Approaching marine conditions are heralded by rising frequencies of coastal taxa such as Chenopodiaceae and *Calystegia*.   Point 16 is an assay on *Phragmites* peat from the base of a biogenic layer, immediately above the silty clay, and the recorded altitude of the isolation contact has also not been greatly affected by consolidation.   The removal of marine conditions, consequent upon the fall of sea-level, were probably rapid: open habitat coastal taxa are represented only by single grains of Chenopodiaceae and *Artemisia*, and high grass pollen frequencies are probably contributed by *Phragmites*.   Low freshwater, aquatic frequencies of *Nymphaea*, *Typha angustifolia*, *T. latifolia* and *Nuphar*, are recorded before a carr vegetation dominated by *Alnus* continues the progressive vegetational development of the site.   The relative fall of sea-level recorded by Point 16 is supported by data from Cartmel, where an assay from Ellerside Moss, at a recorded altitude of +3.72m. O.D. and little affected by subsequent consolidation, yielded a date of 5435+105 (Hv. 3844), and, at Helsington Moss there is a date of 5277+120 (Q. 85) for the end of marine conditions at a recorded altitude of +4.88m. O.D.

Point 17. Lytham Common-1. LC-1. 5005+65 (Hv. 3845) +3.09m. O.D. (Table 1. No. 40).
A fall of sea-level, after a maximum altitude close to +2.94m. O.D. is recorded by Point 17.   The dated sample was a 5 cm. thick slice of *Phragmites*-rich peat from the base of a biogenic layer (Stratum 2) immediately above a marine clay: consolidation has been minimal, and the recorded altitude of the isolation contact must be close to the actual altitude of sea-level at the beginning of Flandrian III.   The pollen diagram (Fig. 28) shows a progressive removal of marine conditions and their replacement by freshwater taxa.   High grass pollen frequencies, probably contributed by *Phragmites*, are supplemented by low frequencies of *Plantago maritima*, *Artemisia* Chenopodiaceae at the base of Stratum 2, and these are replaced by freshwater taxa such as *Typha angustifolia*, *T. latifolia*, *Hydrocotyle vulgaris*, *Potamogeton* and *Alisma*.
              There is consistent evidence for the progressive fall of sea-level during the first few centuries of Flandrian III, and the removal of marine conditions from extensive areas in north-west England, consequent upon this fall.   In Lytham Hall Park, LHP-4 (Table 1. No. 23) records a fall in sea-level at a recorded altitude of +3.39m. O.D., at 4960+210 (Hv. 2919). Elsewhere on Lytham Common, Heyhouses Lane (Table 1. No. 56) records the end of marine conditions at a recorded altitude of +2.87m. O.D., at 4895+95 (Hv. 4344).   North of the River Wyre, the end of marine conditions are dated at 4830+140 (Hv. 4347) at Moss Farm-6 at a recorded altitude of +2.95m. O.D., and at Lousanna at 4900+450 (Hv. 3052) at a recorded altitude of +4.80m. O.D.   Further north in Lonsdale and Cumbria, the fall of sea-level is recorded at the Duddon estuary (Table 1. No. 36), where the isolation contact occurs at an altitude of +4.06m. O.D. at 4960+50 (Hv. 3841).   At Arnside Moss(Table 1. No. 32) it is recorded at an altitude of +4.98m. O.D., at 5015+100, and at Pelutho, in the Solway Lowlands (Table 1. No. 62) at an altitude of +8.70m. O.D. at 4845+100 (Hv. 4418).

Point 18. Peel-2. P-2. 4800+75 (Hv. 3933) +2.24m. O.D. (Table 1. No. 43)
The minimum altitude to which sea-level fell is given by Point 18, which is an assay on woody detrital peat immediately above marine clay.   The pollen assemblage (Fig. 20) shows the progressive removal of marine conditions, with declining frequencies of Chenopodiaceae.
The sea-level minimum is supported by evidence from North Wales, where at Rhyl Beach-1 (Table 1. No. 60), there is a date of 4725+65 (Hv. 4348) for the end of marine conditions at a recorded altitude of +2.43m. O.D.   Consolidation would have been minimal at both these sites, for the assay material came from biogenic sediments immediately above clay with low consolidation characteristics.   At Rhyl Beach a salt marsh clay, with sandy and

ferruginous partings underlies the woody, detrital peat, and the pollen
assemblage shows a rapid replacement of open habitat taxa, such as Gramineae,
Filipendula, *Potentilla*, Chenopodiaceae, *Artemisia* and *Calystegia*, by an
Alder carr.

Point 19. Altmouth. AM-1. 4545+90 (Hv. 2679) +3.14m. O.D. (Table 1. No. 11).
A rise of sea-level, after the nadir recorded by Point 18, is given by
Point 19.   The assay sample comprised an 8 cm. thick slice of *Phragmites*
peat from the base of Stratum 6, immediately above a marine clay.   The
recorded altitude of the isolation contact is +3.11m. O.D., and little
consolidation is anticipated, even though within the biogenic deposit
elliptical cross-sections of *Betula* and *Alnus* testify to autocompaction
within the fen peat.   The pollen assemblage from Stratum 6 shows a
progressive vegetation succession, from open habitat coastal taxa, such as
Plantago maritima, Chenopodiaceae and *Armeria* with high Gramineae
frequencies, undoubtedly contributed by *Phragmites*, a freshwater aquatic
stage in which *Typha angustifolia*, *T. latifolia*, and *Potamogeton* were
represented to an *Alnus-Salix* carr with *Frangula*, *Lythrum salicaria* and
woody climbers such as *Solanum dulcamara*.   Additional data are given elsewhere
(Tooley 1970, 1977).   The fall of sea-level recorded by Point 19 may be
supported from southern Cumbria by data from High Foulshaw Moss (Table 1.
No. 6) where there is a date of 4616+112 (Q.88) at a recorded altitude of
+5.18m. O.D. for the end of marine conditions.

Point 20. Reeds Lane, Moreton. 3965+110 (Q.620). +3.05m. O.D. (Table 1.
No. 9).
There is no evidence for the minimum height to which sea-level fell after
Point 19 was registered, but there can be little doubt from the stratigraphic
and palynological investigations on the Wirral (Godwin and Willis, 1962:
Reade, 1871: Erdtman, 1928: Travis, 1926) that shortly after 3965+110
radiocarbon years ago sea-level again rose, inundated terrestrial biogenic
sediments south of Dove Point, and an estuarine clay was deposited, with
entire valves of *Scrobicularia plana* recorded in their living positions
in the clay.   There is some indirect, supporting evidence from the coastal
sand dunes of West Derby.   At the seaward margin of Downholland Moss,
monocotyledonous peat has been overblown by sand: an assay on peat
immediately beneath the sand yielded a date of 4090+170 (Hv. 4705) at a
recorded altitude of +3.42m. O.D.   If the argument of Jelgersma et al.
(1970) is accepted, that dune instability is closely related to marine
regressions and periods of relatively low sea-level, then both these dates
from different palaeoenvironments are consistent and both indicate a marine
regression, followed by a transgression, the culminating stage of which is
recorded by Point 21.

Point 21. Lytham Hall Park-8. LHP-8. 3150+150 (Hv. 2918) +3.51m. O.D.
(Table 1. No. 22).
The altitude to which sea-level rose prior to the registration of Point 21
has been estimated.   There is unequivocal evidence for the end of marine
conditions at LHP-8, at a recorded altitude, little affected by consolidation
of 3.51m. O.D.

Point 22. Lytham Hall Park-8. LHP-8. 3090+135 (Hv. 2917). +3.70m. O.D.
(Table 1. No. 21).
A subsequent rise of sea-level is recorded by Point 22, after a short-lived
period of relatively low sea-level.   Approaching marine conditions are
indicated by low persistent frequencies of Chenopodiaceae, *Calystegia cf.
soldanella*, *Cynoglossum* and *Armeria*, replacing freshwater aquatic taxa,
such as *Hydrocotyle*, *Iris* and *Typha angustifolia*.   The altitude to which
sea-level rose has been estimated, but that it was still underway after 2820
radiocarbon years ago is borne out by evidence from the Annas Mouth in
Cumbria (Table 1. No. 37).   Here, at a recorded altitude of +6.69m. O.D.,
a humified, monocotyledonous peat gives way to a brackish water clay 74 cm.

thick.   An assay on the peat immediately subjacent to the clay yielded a date of 2820+55 (Hv. 3842).

Point 23. Lytham Hall Park-8. LHP-8. 2270+65 (Hv. 2916) +4.22m. O.D. (Table 1. No. 20).
Point 23 records the end of marine conditions, and a relative fall of sea-level.   The assay sample was a gyttja with *Phragmites*, immediately above a coarse silt of marine origin: the progressive removal of marine conditions is indicated in the pollen diagram, by the reduction of frequency of open habitat, coastal taxa, such as Chenopodiaceae and *Artemisia*, and the development of a *Salix* carr.   There is supporting evidence from Formby Foreshore (Table 1. No. 67), where a dune slack deposit ceased accumulating shortly after 2335-20 (Hv. 4709) at an altitude of +5.08m O.D., and there can be little doubt that the burial of the dune slack by sand is associated with a relative fall of sea-level, recorded by Point 23.

Point 24. Lytham Hall Park-1. LHP-1. 1800 (estimated) +5.44m. O.D.
There is an estimated age for Point 24 which is the highest marine deposit recorded in west Lancashire: the estimation is based on both altitude and pollen assemblage considerations, and the relationships of the index point to Points 23 and 25.   A clay of estuarine origin underlies a layer (Stratum 3) of highly compressed, woody detrital peat.   The pollen assemblage (Fig. 31) is dominated by Gramineae and Chenopodiaceae, and there are low frequencies of *Plantago maritima*, *Litorella*, *Artemisia*, *Ruppia*, and *Rumex*.   Above the index level, a progressive vegetation development to freshwater communities is indicated by high frequencies of *Typha latifolia* and *T. angustifolia*, together with low frequencies of *Iris* and *Potamogeton*. Support is given to the estimated age of 1800 radiocarbon years ago by the date from Ansdell-2.   Here, an assay on monocotyledonous peat with *Phragmites* immediately above sandy silt of marine origin yielded a date of 1795+240 (Hv. 5215), at a recorded altitude for the isolation contact of +4.05m. O.D. The declining frequencies of open habitat, coastal taxa, such as Chenopodiaceae, *Artemisia* and *Silene* and the increase in frequencies of freshwater taxa, such as *Potamogeton*, *Iris*, *Hydrocotyle*, *Typha angustifolia*, *T. latifolia* and *Lemna* through the peat undoubtedly reflect the progressive removal of marine conditions, consequent upon a relative fall of sea-level.

Point 25. Ansdell-1. A-1. 1370+85 (Hv. 4708). +4.60m. O.D. (Table 1. No. 66).
The continued removal of marine conditions sustained until after 1370 radiocarbon years ago is indicated by this index point, and a low sea-level is inferred on the basis that the inception of sand drifting shortly after 1370 radiocarbon years ago is associated with a marine regression.

Point 26. Lytham Common-1. LC-1. 805+70 (Hv. 4417). +5.63m. O.D. (Table 1. No. 61).
There is some indirect evidence for a period of relatively high sea-level during the Dark Ages.   The sand dunes along the Fylde coast are subdivided stratigraphically by a well-developed biogenic stratum of monocotyledonous peat which ranges in age from 805+70 (Hv. 4417) to 830+50 (Hv. 3846). Point 26 refers to the former date.   The pollen assemblage from the assay level (Fig. 28) is dominated by dwarf shrubs, such as *Calluna* and by Gramineae, Cyperaceae, *Sphagnum* and Filicales.   *Corylus* frequencies are high and immediately subjacent to the overlying stratum of sand, *Populus* values rise.   It is probable that periods of dune instability and sand drifting are associated with periods of relatively low sea-level (Jelgersma et al. 1970) whereas periods of dune stability, wet slacks and peat accumulation are associated with relatively high sea-levels.   The biogenic stratum that accumulated in the Starr Hills, Lytham between 830 and 805 radiocarbon years ago is undoubtedly closely associated with a period of relatively high sea-level.   The altitude attained is conjectural, and cannot be deduced from the altitudes of the fossil dune slacks, unaffected directly by the marine transgression.

THE HILLHOUSE COASTLINE

The establishment of a sea-level curve from a limited area in west Lancashire showing periods of relatively high and low sea-level and associated marine transgressions over the last 9000 years has important implications for a consideration of the age and nature of the *Hillhouse coastline*, defined by Gresswell in 1953 (see Tooley 1976b).

Gresswell explained the evolution of the coast in north-west England in terms of a single marine transgression, consequent upon a relative rise of sea-level to an altitude of at least +5.79 m O.D. Such an elevation was necessary to erode a cliff in the till at a mean recorded altitude of +5.18 m O.D. In 1953, Gresswell argued that the period of relatively high sea-level in south-west Lancashire occurred about 3000 B.C. (5000 radiocarbon years ago), whereas in 1958 he was able to conclude on the basis of pollen analytical evidence (Oldfield 1958) that the period of relatively high sea-level in southern Cumbria was confined to the Boreal-Atlantic transition, that is 7100 radiocarbon years ago. In 1967, he was convinced that the period of relatively high sea-level occurred 6000 radiocarbon years ago. To the cliff cut in the till, Gresswell gave the name *Hillhouse coastline* and to the dilated Irish Sea, that pushed landward of the present coast, the *Hillhouse Sea*. The type locality for the *Hillhouse coastline* was the area immediately west of Hillhouse Farm (Fig. 11), which Gresswell described in 1953 (p. 28):

[At Hillhouse, an outcrop of Keuper Sandstone] rises straight out of the peat of Altcar Moss, present surface about eight feet O.D., with the boulder-clay surface certainly well below three feet. The rock rises to thirty-seven feet, and still forms a fairly steep slope on its western and south-western sides . . . Hillhouse therefore provides a very good feature so far as the coastline is concerned, and, clearly, as would be expected, the outcrop formed a promontory of more resistant material, with a tail of uneroded clay behind. This locality has been used to give a name to the whole length of this coastline.

A relative fall of sea-level to a nadir of -27.0 feet O.D. (-8.23 m O.D.) after the *Hillhouse coastline* had been formed is explained by Gresswell (1953, p. 46) as the result of a diminished or terminated eustatic rise of sea-level, and continued isostatic uplift. To this period of relatively low sea-level, Gresswell assigns the deposition of firstly Shirdley Hill Sand and secondly Downholland Silt, on a prograding shoreline.

In order to establish the relationship of the *Hillhouse coastline* to the pattern of marine transgressions and the fluctuating sea-levels upon which they were consequent, detailed stratigraphic and pollen analyses were carried out at the type locality. The results of these analyses are presented in Chapter 2.

On Fig. 11, the *Hillhouse coastline* has been plotted according to the plan given by Gresswell (1953, Fig. 10(B), p. 27) for the type locality. The 25 ft. (7.6 m) contour has also been drawn, and this clearly shows the degree to which the *Hillhouse coastline* is height transgressive in this area. Sampling sites HH-1 to HH-6, shown on Fig. 11, were located at points where a slope inflexion had been identified in the field: HH-3 was located precisely on the line of the *Hillhouse coastline*. The orientation of the line of sampling sites north-east to south-west was on the basis that, if the feature·was a degraded cliff-line, then it would have been developed best along that part of the promontory facing the approach of storm waves from the south-west. Figure 12 shows the stratigraphic succession across the *Hillhouse coastline*. The stratigraphy at H-2 and -3 clearly shows a boulder clay passing up into a weathered sand, and at HH-4 and -5 this passes up to a woody detrital peat. At HH-6, biogenic sedimentation is interrupted by a layer (Stratum 4) of blue-grey, silty clay with altitudinal limits of -1.12 to -0.57 m O.D. A pollen diagram from an adjacent site, HH-6A (Fig. 13) shows a modest rise of fresh-water taxa such as *Typha angustifolia* in the biogenic layer (Stratum 2) replaced by open habitat, coastal taxa, such as Chenopodiaceae and *Plantago maritima*, immediately subjacent to the clay. Rising grass frequencies are probably contributed by *Phragmites*. The end of the deposition of the clay and a progressive vegetation development are indicated in the pol-len assemblage zone by declining frequencies of Chenopodiaceae, rising frequencies of freshwater taxa, such as *Typha angustifolia*, and by rising *Alnus* frequencies. There can be little doubt, that the layer of blue-grey, silty clay is marine in origin, and that marine conditions commenced at an altitude of -1.12 m O.D. and ended at -0.57 m O.D. at HH-6, and penetrated land-ward little further than HH-6, for between HH-6 and HH-5, the subsurface of coarse sand rises from an altitude of -1.27 m O.D. to +1.18 m O.D..

The pollen assemblage from HH-6A is undoubtedly early Flandrian II, and the marine transgression recorded here is homologous to the transgression DM-I recorded further north on Downholland Moss at a mean altitude of -0.72 to -0.19 m O.D. There is no evidence from the landward margin of Altcar Moss, adjacent to the *Hillhouse coastline*, that more recent transgressions affected this area; indeed, during the time and altitudes covered by biogenic sedimentation at HH-6, another two transgressions were recorded in West Derby.

In order to reconcile the evidence presented here with the position of the *Hillhouse coastline* shown on Figs. 11 and 12 it would be necessary to evoke a mechanism by which storm waves rose more than six metres from a quiet water, sedimentary environment, in which brackish water clays were accumulating, to cut a notch in the till at Hillhouse, at least a thousand years after marine sedimentation ended at HH-6. Such a mechanism is wholly untenable, and the requirements for erosion at HH-3 contradict those for deposition at HH-6, separated by only 600 metres.

Further north, on Downholland Moss, Gresswell's *Hillhouse coastline* has been plotted on Fig. 5, to show its relationship to the sampling sites on Downholland Moss, and to the eastern limit of the grey clays mapped and described by de Rance (1868, 1869). The maximum landward penetration of the sea between 6980 and 6760 radiocarbon years ago at recorded altitudes at

DM-11A of -0.36 to -0.16 m O.D. is clearly shown on Figures 6 and 33.

In order to reconcile these data with the position and assumed age of the *Hillhouse coastline* east of Downholland Moss, it is necessary to explain how sea-level could have risen another 6 metres above the recorded altitudes of the DM-I transgression, and penetrated landward 1.3 km more than a thousand years after the last marine transgression recorded the furthest to the east in this area during the Flandrian stage.

In the Martin Mere area, a complete stratigraphic survey of the Mere sediments and subsequent pollen analysis (Tooley 1977a) proved that there were no marine sediments south and east of Midge Hall in Martin Mere basin. Again it is difficult to explain how sea-level could rise, inundate the whole of the Mere basin and cut a notch in the till, which at Holmeswood has an altitude of +5.18 m O.D., at a time when freshwater reedswamps, with a vegetation mosaic of *Typha angustifolia* and *Cladium mariscus* occupied areas marginal to the open water of the Mere. Indeed, it is probable that the *Hillhouse coastline* around Martin Mere represents a former lake shore.

Further north, in southern Cumbria, such an interpretation is more rational for the feature that Gresswell mapped as a raised beach in the Winster Valley, and which, from the evidence Smith (1958b) presents, can be explained as a lake shore formed by an enlarged Helton Tarn. In the Rusland valley, Dickinson (1973) presents incontrovertible evidence and states explicitly that there was no marine transgression beyond Crooks Bridge in the Rusland valley, and that about 6000 radiocarbon years ago a freshwater lake occupied that part of the Rusland valley now covered by Rusland Moss, Hulleter Moss and Hay Bridge Moss. Furthermore, the limit of the raised beach shown in the Rusland valley by Gresswell (1958) coincides very nearly with the shoreline of the freshwater lake deduced from stratigraphic evidence by Pearsall in 1918.

Finally, a consideration of the altitudes on Gresswell's *Hillhouse coastline* shows that in south-west Lancashire the *coastline* has an altitude of +5.18 m O.D. (Gresswell 1953). This compares with an altitude of +4.65 m O.D. for raised beaches of southern Cumbria (Gresswell 1958). This negative movement between south and north Lancashire, Gresswell explains (1958, p. 98) by the points of measurement. In south west Lancashire:

the level was measured at the actual toe of the boulder clay cliffs. The beach shelves away immediately seaward of the cliff-line, and the figure of 17 feet O.D. is thus the limit of spring tides, the limit of the erosion bench of storm waves. On the other hand, the infilled estuaries in Furness and Lyth are areas of deposition, and would be about three feet awash at full tide. A silt level of over 15 feet thus brings the high water level to over 18 feet, and in fact the wave cut bench on Winder Moor is at 19 feet.

Gresswell had argued earlier (1953, p. 46) that, for erosion to take place along the *Hillhouse coastline* in south-west Lancashire, sea-level here must have been close to 19 feet (5.8 m) O.D. Nevertheless, the correction factor of 3 feet was added to the altitudes from southern Cumbrian beaches, and a figure of 1 to 2 feet was derived for the differential amount of isostatic recovery between the two areas. Without numerical transformation of the measured altitudes to permit a rational explanation of the observed *fall* in beach heights from south to north, the height data for the raised beaches are clearly ambiguous.

The stratigraphic, radiometric and pollen analytical evidence presented here, tends to contradict both Gresswell's evidence and interpretation, and it is probable that the *Hillhouse coastline* is a fiction.

THE EVOLUTION OF THE LANCASHIRE COAST

The stages in the evolution of the Lancashire coast are intimately related to the relative changes of sea-level. The rise of sea-level between 9200 and 5000 years BP was registered by extensive transgressions of the coast. At the opening of Flandrian III (5000 years BP), relative sea-level on the Lancashire coast was one metre below its present level, and, although sea-level continued to rise some 2.5 metres in the succeeding 4000 years, eroding till headlands, most of the constructional landforms, such as the shingle spits, sand-bars and sand-dunes date from this period.

The following landforms are recognised on the Lancashire coast, and have been affected by the rise in sea-level along the coast, although they are not genetically related in every case:

1. *Solid outcrops* are characteristic of Lonsdale. Elsewhere the solid is mantled by drift and more recent unconsolidated deposits, and is proved only in borings. Bunter Sandstone forms a low, westward-facing cliff at Cockersand, and around Morecambe Bay, headlands of Carboniferous Limestone are separated by valleys filled with drift and Flandrian marine sediments.

2. *Till* of Devensian age forms cliffs in Lonsdale and Amounderness, although in the latter case these are largely obscured by revetments and coastal protection works. Till sections are well exposed on Walney Island, around Morecambe Bay and in the Wyre estuary. In south-west Lancashire, the till is obscured by Flandrian sediments, but is at, or close to, the surface east of Crossens. Elsewhere, it is proved in borings and temporary sections. Lancashire is the type area for the tripartite sequence of Upper Boulder Clay, Middle Sands and Lower Boulder Clay. This sequence is attributed to a single complex glaciation between 25000 and 20000 years BP followed by oscillations of the ice front during a general melt. There are differences of opinion about the chronology of the oscillations and the correlation of re-advance limits (Huddart *et al.* 1977, Tooley 1977a).

3. *Shirdley Hill Sands.* This formation is recorded largely from south-west Lancashire, although an outlier is described from the Fylde by Hall and Folland (1970). The sand rarely exceeds a thickness of 5 metres and has been recorded from -14 metres O.D. beneath Downholland Moss to +120 metres O.D. on the western slopes of Billinge Hill. West of Haskayne, there are low parabolic dunes, but such distinct landforms are not common throughout its range. The sand is intercalated with unhumified woody

detrital peat, sandy monocotyledonous peat and organic sand. Horizons of bleached, white sand beneath the latter and weakly developed illuvial horizons containing ferruginous nodules indicate that podsolic palaeosols developed extensively at different stages throughout the formation. In the vicinity of Ormskirk, at Clieves Hills, depressions in the till reveal the following succession: a till unit overlaid by head and in turn by an unhumified peat, which has been buried by Shirdley Hill Sand with organic horizons intercalated. Microfossil and radiometric analyses of the biogenic sediments indicate that the first period of sand-drifting began here shortly after 10455±110 years BP (Table 1, No. 68), that is during LDeIII, when many of the Lake District corries were re-occupied by ice (Pennington 1970) consequent upon a climatic deterioration. In south-west Lancashire, the Shirdley Hill Sands appear to have remained unvegetated and unstable until Flandrian Ib, when stability and a more or less continuous vegetation cover of birch and pine resulted in biogenic accumulation within depressions in these inland dunes. Locally, in the area of Firswood Road, the dunes remained unstable until the middle of Flandrian II. The dunes became generally unstable towards the end of Flandrian II and during Flandrian III as a direct result of clearance and deforestation; indeed, the more open nature of woodland in these areas must have made them more attractive to clearance during the late Mesolithic and Neolithic periods than the alder-oak fens and reedswamps of the perimarine and lagoonal zones below +7.0 metres O.D. There is little doubt that the Shirdley Hill Sands are cover sands, which accumulated during the culminating stages of the Devensian glaciation, before the Flandrian rise of sea-level brought marine facies to the Lancashire coast during Flandrian Ib, the transgression of the coast at this time, the reduction in continentality and an increase in humidity were factors which contributed to the stability of the formation rather than its cause (see also Tooley and Kear 1977).

4. *Perimarine Zone.* This zone has been affected indirectly by the rise of sea-level along the Lancashire coast. It comprises a narrow belt of land between the tidal flat and lagoonal zone to seaward, affected directly by marine transgressions, and the rising till surfaces to landward. The sediments of the perimarine zone are freshwater clays and silts, gyttjas, reedswamp and terrestrial peats. The extent of the perimarine zone has not been established throughout Lancashire. In south-west Lancashire, it comprises Downholland Moss between sampling sites DM-12 to DM-14, at Hillhouse between HH-3 and HH-5 and the majority of Martin Mere during Flandrian II and III (Tooley 1977a, and unpublished). In Lonsdale the whole of Carnforth Levels can be defined as a perimarine zone.

The perimarine zone contracts and dilates depending on the extent of marine transgressions, registered in the tidal flat and lagoonal zones. Thus, after Lytham VI, the perimarine zone underwent considerable dilation, for no subsequent transgressions were as extensive as Lytham I–VI and were confined to embayments of limited extent, yet the effect of these transgressions was a regional elevation of the ground-water table. Hence, extensive areas of Lytham Common and Lytham Moss in the Fylde, Pilling, Cockerham and Winmarleigh Mosses in Over-Wyre, and the interconnected series of mosses from Martin Mere to Altcar Moss, comprised the perimarine zone from about 4800 years BP onwards. Similarly, Heysham Moss, Silverdale Moss, Foulshaw Moss, the Ellerside–White Moss–Deer Dike Moss complex along the eastern margin of the Leven Estuary, and the Waitham Common–White Moss complex in the valley of Kirkby Pool constitute the perimarine zone of Flandrian III in Lancashire and southern Cumbria.

5. *Tidal Flat and Lagoonal Zones.* Within this area marine transgression sequences have been established. They comprise the mosslands, seaward of the perimarine zone and pass beneath the coastal sand-dunes: the stratigraphy of the mosslands in this area consists of alternating strata of marine clays, silts and sands and brackish to freshwater biogenic deposits. The freshwater deposits are invariably gyttjas, in which aquatic taxa are recorded. In south-west Lancashire, Black and White Otter, Gettern Mere, Martin Mere and Downholland Mere represent lagoonal stages; the mere, of similar origin, on Lytham Moss is Cursidmere. In the Fylde, the tidal flat and lagoonal deposits span the period from 9270 until 805 years BP, during which time ten marine transgressions affected the Lancashire coast. These transgressions have the following time limits (years BP):

Lytham I	9270–8575
Lytham II	8390–7800
Lytham III	7605–7200
Lytham IV	6710–6157
Lytham V	5947–5775
Lytham VI	5570–4897
Lytham VII	3700–3150
Lytham VIII	3090–2270
Lytham IX	1795–1370
Lytham X	~ 817

The inception of lagoonal stages is not synchronous: Martin Mere in West Derby became a freshwater lake at the beginning of FII, although its seaward margin near Wyke House Farm was affected by subsequent marine transgressions. Marton Mere in the Fylde was unaffected by sea-level changes and its sediments contain a record of Late Devensian and Flandrian environmental changes. Downholland, Black and White Otter and Gettern Meres date from the opening of Flandrian III.

6. *Sand-dunes, sandbanks and shingle spits.* During the relative rise of sea-level recorded along the Lancashire coast, there was little opportunity for these features to develop. Any incipient development during periods of low

sea-level was overtaken by a successive rise in sea-level, and the removal of unconsolidated material either to the Irish Sea basins or to areas suitable for accumulation. After 4000 years BP, with an oscillating sea-level close to or slightly above the present level, longshore drifting was maintained between limits, similar to those recorded at present along the coast. During periods of relatively low sea-level about 4200, 3100, 2200 and 1200 years BP the potential for sand accumulation and dune instability was increased, whereas in succeeding periods, during which marine transgressions were under way, an increase in humidity and water levels in the dunes would result in greater stability. In addition, there was abundant unconsolidated material available of different grade sizes from the till cliffs and fluvio-glacial deposits, which was re-worked between tidal limits, or, in the case of Morecambe Bay, introduced by tidal currents.

A precise sand-dune chronology has not yet been established in Lancashire, but the following framework is proposed. At the western margin of Downholland Moss, peat growth was ended during the first millennium of Flandrian III: a radiocarbon assay on *Phragmites* peat immediately subjacent to the sand gave a date of 4090±170 years BP (Table 1, No. 63). This maximum landward penetration of sand may be related to Lytham VIa: an oscillation of relative sea-level at this time, carrying mean sea-level down to -2.3 metres O.D., would increase the effective intertidal area from which sand could be winnowed. In addition, the freshwater table within the sand-dunes may have become depressed, and concomitant erosion resulted.

In the Netherlands, the oldest period of dune building, Oldest Dunes (O.D.A.) was completed before 4100 years BP, and occurred whilst a trans-gression, Calais IV, was under way (Jelgersma *et al.* 1970). The dunes of the south-west Fylde coast are not as ancient as those in south-west Lancashire, whereas on Walney Island, the dunes at the North End bury occupation layers of Neolithic age, and are of similar age to the Formby and Ainsdale Hills in Lancashire.

During the two succeeding millennia, there is no evidence from the dune systems either for dune building or for dune stability: the occurrence of two marine transgressions, Lytham VII and VIII, indicates that two cycles of stability/instability should have been registered. On Formby Foreshore, a date on peat from the base of a peat bed, immediately overlying a white, bleached, coarse sand, was 2335±120 (Table 1, No. 67): it is probable that this date records a period of relative dune stability. In the Netherlands, a dune-building period, Older Dunes IIb, was under way, and at Velsen-Hoogovens IV there is a date of 2450±40 on *Juniperus* wood, embedded in a sand layer, referable to this period.

In the dunes of the south-west Fylde, there is a well-marked peat horizon, intercalating the blown sand; the sand recorded throughout Lytham Common can be subdivided on the basis of this horizon. At LC-1, the peat is 4cm thick and at LHP-6 it is 12 cm thick: an assay on the former gave a date of 805±70 years BP (Table 1, No. 61) and on the latter, 830±50 years BP (Table 1, No. 41). The sample was taken from the middle of each peat horizon to minimise the chance of contamination from base rich waters. At some stage before and after the absolute date, a period of dune building ended, and another began. These periods can be assigned with some confidence to the second phase of Younger Dunes identified in the Netherlands, Y.D.Ib, and assigned to the twelfth to thirteenth centuries A.D. (Jelgersma *et al.* 1970). There is, furthermore, documentary evidence from the chartulary of Lytham Priory of dune instability at this time, and in the latter half of the fifteenth century (Fishwick 1909; A. Piper, pers. comm.).

This preliminary, chronological framework for the sand-dune systems of the Lancashire coast, together with their probable correlatives in the Netherlands, encourages further detailed examination of dune systems. The establishment of a sequence of dune-building stages, separated by periods of dune stability, correlated across north-west Europe, will have important implications for the interpretation of evidence of climatic change during the last 4000 years.

7. *Salt Marshes.* The present distribution of saltmarshes on the Lancashire coast reflects the most recent stage of relative sea-level changes in this area. Extensive salt marshes are found in the Mersey and Ribble estuaries and around the margins of Morecambe Bay. Salt marshes of limited extent are found at the mouth of the River Alt, within the River Wyre, River Lune and Duddon estuaries. Of these, only the mouths of the River Alt and River Winster carry ungrazed salt marshes. The area of salt marsh has contracted over the last 200 years particularly as a result of enclosure and reclamation. A full description of saltmarsh development in the Ribble estuary can be found in Barron (1938), Gresswell (1953), Berry (1967) and Cheesbrough *et al.* (1969). The dilation of the Lancashire coast salt marshes, which has occurred since 1950, when *Spartina townsendii* began to colonise the inter-tidal zone of the Ribble estuary, has not approached the area under salt-marshes, indicated by each marine transgression recorded during earlier episodes in the Flandrian stage.

Of the seven distinct landforms recognised on the Lancashire coast, the tidal flat, lagoonal and perimarine zones, the sand-dunes, shingle spits, offshore bars and salt marshes are inextricably linked in their evolution to the changes in sea-level that have affected this coast. The present stage of slope development on coastal units backed by till cliffs reflects the most recent adjustments of relative sea-level, whereas the notch cut in limestone cliffs around Morecambe Bay may reflect in part relatively higher sea-levels from 3500 to 800 years BP, and in part inheritance, during the Flandrian stage, of older erosional features resulting from higher sea-levels during the Ipswichian interglacial, some 8 to 13 metres above present sea-level (Sparks and West

1972). Such an interpretation is proposed for the weathered, concave notch at the base of the limestone cliffs at Silverdale. Seacaves and notches at higher altitudes (Ashmead 1974) may be of Hoxnian interglacial age, or an earlier temperate stage during the Quaternary.

The following stages in the evolution of the Lancashire coast can be given. They are based on detailed systematic stratigraphic, microfossil and radiometric analyses from Downholland Moss in south-west Lancashire, Nancy's Bay, Lytham Moss, Lytham Hall Park and Lytham Common in the south-west Fylde, and are corroborated by similar analyses at other sites throughout Lancashire. A palaeogeography of Flandrian marine deposits can be suggested from interpolation of these sites.

1. The position of mean sea-level in Lancashire 9000 years BP was about -21 metres O.D. The central and southern marginal parts of Morecambe Bay had been inundated by rising sea-level, but the till subsurface beneath the present coasts of south-west Lancashire and the south-west Fylde was too high (at -11.0 metres O.D.) to permit the registration of this early transgression (Lytham I). The evidence probably resides offshore (Tooley 1977a; Pantin 1977). A later stage of Lytham I was registered in these areas, and, in the Fylde, during the culminating period of the transgression, marine facies had been carried 2.2 km landward of the present coast. Downholland Moss had also been transgressed, at least as far as DM-5.

2. A fall of sea-level between Lytham I and Lytham II exposed the former intertidal zone to weathering, and locally, biogenic sedimentation began within depressions. In favoured locations limnic and monocotyledonous peats accumulated for 700 years and more, and ended when sea-level rose and inundated these areas. Immediately before inundation, at an altitude of -11.13 metres O.D. beneath the Starr Hills at Lytham, the presence of dense layers of charcoal (dated at 8390±105 years BP), declining frequencies of tree taxa, an increase in the frequency of *Corylus* and of open habitat taxa, indicate interference with the vegetation in this area by Mesolithic folk.

3. During a subsequent rise of sea-level, marine sediments were deposited, and can be ascribed to Lytham II. In the south-west Fylde, Lytham II transgressed the present coast, shortly after 8400 years BP, and had penetrated more than 2 km landwards by 7800 years BP, so that during Flandrian Id, the coast of the south-west Fylde had been pushed landwards from its present position to a line close to Heyhouses Lane (Fig. 24). Its maximum landward penetration can only be approximated. The till subsurface was rising both northwards and eastwards, and landward penetration by Lytham II was probably confined by a line close to the landward limit of blown sand, some 3.0 km from the present coast. The present coast of Morecambe Bay had been transgressed, and the lower valleys of the rivers Lune, Kent, Winster and Leven inundated. Barrow Harbour was also

inundated, and Walney Island became separated from the mainland during Lytham II, if not during Lytham I. In south-west Lancashire, Downholland Moss probably suffered a second marine transgression, and the sea impinged on the northern margin of Martin Mere (Tooley 1977a).

4. Lytham III and Lytham IV were extensive marine transgressions landward of the present coast. On Downholland Moss, the maximum landward penetration occurred shortly after 6980±55 years BP, immediately east of DM-11. Lytham IV was prevented from inundating Martin Mere by the sandbank which accumulated during Lytham III and the transgressive phase of Lytham IV immediately south of Churchtown Moss. During the transgressive phase of Lytham IV, the Lytham–Skippool valley was inundated, the Wyre and Ribble estuaries linked by way of this valley, and the western Fylde became an island.

5. Lytham V registered a slight rise in relative sea-level and local dilation of estuarine conditions in Downholland Moss, Nancy's Bay and Silverdale Moss. Lytham VI affected a much more extensive area, and the subsequent removal of marine conditions during the first two hundred years of Flandrian III is a well-marked event in Lancashire. The tidal flat zone of south-west Lancashire, running from the Gravel near Crossens to the Alt Mouth, was evacuated, and no subsequent marine episode was recorded directly throughout this area. In the Fylde, Lytham Moss and Brown Moss also recorded extensive removal of marine conditions and the inception of biogenic sedimentation, uninterrupted by subsequent marine transgressions. After 2000 years, the east and west Fylde were reunited, and biogenic sedimentation within the Lytham–Skippool valley prevented further marine transgression. In Lonsdale, areas adjacent to the River Lune ceased to register marine conditions, and in the valleys of the rivers Gilpin, Kent, Winster and Leven, estuarine conditions gave way to telmatic and terrestrial conditions before ombrogenous bog developed extensively. Radiocarbon dates on the transition from estuarine to terrestrial conditions around Morecambe Bay indicate that the date was advanced by several hundred years, and is undoubtedly the result of isostatic recovery in this area. However, in the Duddon estuary, removal of marine conditions was delayed until the opening of Flandrian III, and falls well within the time limits established for south-west Lancashire and the Fylde for Lytham VI. Inlets at the Alt Mouth, Formby, Churchtown, Hesketh Bank, Nancy's Bay, Fairhaven, St. Nicholas Lane and Arnside Moss became progressively silted up, and in the case of Formby and Churchtown buried by blown sand after 4000 years BP. Other inlets remained open, draining the mosslands and, as vulnerable areas on the coast, serving as points of ingress by the sea during periods of relatively high sea-level, recorded in subsequent millennia.

6. Lytham VII to X were confined progressively to inlets adjacent to the estuaries of the Lancashire coast. Further extension of marine conditions was prevented by the reduced rate of sea-level rise, and by an accelerated

rate of biogenic sedimentation, consequent upon the climatic deterioration that occurred about 2500 years BP. The most vulnerable inlet on the north side of the Ribble estuary, west of Church Scar, appears to have given access to Lytham Hall Park, protected at its seaward end by the recurved shingle spit at Fairhaven: Lytham VII to X are all recorded in this area, and a combination of aspect (south-west) and discharges of water from the mosslands to landward of the inlet prevented the accumulation of shingle across the mouth of the inlet until about 1000 years ago.

7. The final episodes on the Lancashire coast have resulted in the development of the salt marshes and sand-dunes. Two dune-building periods are recognised: from 4000 to 2400 years BP, which is correlated with the Older Dunes of the Netherlands, and during the twelfth and thirteenth centuries, when the Younger Dunes were in the process of formation in the Netherlands.

This scheme for the evolution of the Lancashire coast is, both in broad outline and in detail, a contradiction of the scheme proposed by Gresswell in 1953. The coastal landforms that Gresswell described were explained in terms of oscillating land- and sea-levels, with a magnitude since 5000 years BP of 40 feet (12.2m) land uplift in south-west Lancashire. This figure is too great, and derives from a misinterpretation of the field evidence from south-west Lancashire. The landforms described in this section undoubtedly derive from a rising sea-level, and from fluctuations of sea-level resulting from eustatic oscillations, the magnitudes of which were increased or decreased in Lancashire by residual uplift in north Lancashire, and downwarping from hydro-isostasy and sediment consolidation in south Lancashire. Any residual uplift in south Lancashire between 9000 and 5000 years BP, was so slight that each eustatic oscillation was registered at progressively higher levels in the tidal flat and lagoonal zone of south-west Lancashire and the Fylde.

CHAPTER 5

North-West European Correlations

1. FLANDRIAN MARINE TRANSGRESSION SEQUENCES

In Chapter 3, ten marine transgressions have been established from stratigraphic records and altitudinal variations; evidence of their age derives both from radiocarbon dating and from pollen analysis of the intercalated biogenic sediments. In some cases, only a relative chronology exists, but the central position of the Flandrian type site at Red Moss, Horwich, Lancashire, where a chronozone sequence has been established (Hibbert *et al.* 1971) permits some measure of confidence in referring Local Pollen Assemblage Zones to the Regional Pollen Assemblage Zones at Red Moss. In other cases, where· transgressions have been established by altitudinal differences only, correlation can be by reference to the zero datum of the United Kingdom. In yet other cases, where absolute dates are available and have been substantiated by referring the LPAZ to the RPAZ and through this to chronozones, the dates refer to the transgressive and regressive contacts of the transgression at the sampling site alone: at only a few sites have the temporal stages of transgressions and regressions been established, and there can be no guarantee that the dates obtained represent the mean time value of the transgression. A mean time value of the beginning and end of a transgression can only be obtained from a sample population of dates on both the transgressive and regressive contacts. The data are then susceptible to statistical analysis and frequency histograms can be constructed, taking into account the standard error of the absolute dates (see Geyh 1969).

The transgressions recognised in north-west England are shown schematically on Figure 35 and have been correlated with those established in North Wales, north-east England, the Fenlands, the Somerset Levels and North-west Europe in Figs. 39 and 40. The basis for correlation is considered in this chapter.

1.1 LOWLAND BRITAIN

Stratigraphic successions similar to those recorded in the perimarine, tidal flat and lagoonal zones of north-west England and North Wales are found throughout the British Isles; attention is here drawn to this similarity in north-east England, the Fenlands and the Somerset Levels, for which systematic examination and descriptions of the stratigraphy have been published. Emphasis has been given to the stratigraphic record, rather than to an elaborate comparison of altitudes and ages, which can only proceed from a large population of well-dated strata whose relationships to a contemporary sea-level have been established. Where those data exist, a tentative correlation can be made with the temporal pattern of marine transgression sequences in north-west England. Such a correlation, however, is subject to two fundamental difficulties: the time limits bounding transgression sequences in north-west England are based on radiocarbon dates from single profiles and may provide dates that are eccentric to the mean time values of the beginning and end of a transgression: dated strata from other areas in Britain are subject to the same difficulty. Local effects, such as downwarping, sediment consolidation or isostatic recovery may advance or retard or reduce the duration of a transgression within particular areas. Thus, precise correlation is elusive.

North-east England

The coast of north-east England comprises the coasts of Northumberland, Durham, Yorkshire and Lincolnshire, as far south as Chapel Point. Along this coastal unit, the opportunities for quiet water sedimentation are few, and are confined to the estuaries of the Tees and Humber, low-lying valleys tributary to these estuaries and the coastal margins of Lincolnshire.

In Northumberland, Hogg (1972) has described the extent and altitude of raised beaches on Lindisfarne and the Inner Farne, at Whitelee Letch, Ross Links and Dunstanburgh Castle. At the foot of the cliff on Lindisfarne, north-west of the castle, the mean altitude is +5.89 m O.D., and from the sandy raised beach between the foot of the cliff and the present shore at an altitude of +4.67 m O.D., a bulk sample yielded a rich molluscan fauna dominated by *Littorina littorea* (Linné.), *Patella vulgata* Linné and *Litorella littoralis* (Linné) (det. D. Gilbertson), and indicating a rocky environment nearby. At Whitelee Letch, Hogg gives the altitude of the base of the cliff as + 7.01 m O.D., and from the raised beach seaward of the cliff a molluscan assemblage was recorded in which *Mya* spp., *Lutraria* sp., *Scrobicularia plana* da Costa and *Lacuna crassior* (Mont.) were the most abundant and indicated intertidal conditions on a sandy or muddy foreshore. The age of these beaches can only be inferred from the pollen analytical evidence from the Inner Farne, south-east of Middle Pond. Both the stratigraphy and the pollen analysis indicate two periods of marine transgression, the first towards the end of Flandrian II and the second at some time early in Flandrian III (Hogg, op. cit.). If this interpretation is confirmed, there is evidence that both Lytham VI and Lytham VIa are registered on the Northumberland coast.

In Durham, cliffs of Magnesian Limestone, capped by till provide little opportunity for the registration of Flandrian marine transgressions until the estuary of the River Tees is reached. Agar (1954) has described the sediments of the Tees estuary, where, at Middlesbrough, there is a maximum thickness of 25.6 metres of sandy silt of marine origin from -27.4 metres O.D. to -1.8 O.D. overlaid by woody, detrital peat. At an altitude of -7.9 m O.D. on the south side of the Tees estuary a sample of peaty sand yielded a pollen assemblage dominated by *Pinus* and *Corylus* which indicates a Flandrian Ic age: the peaty sand is overlaid by grey silty clay of marine origin. These data show

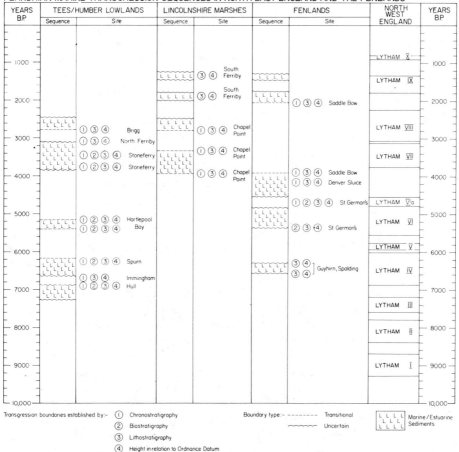

FIG. 39. Scheme of Flandrian Marine Transgression sequences in north-east England and the Fenlands. The evidence from the Tees and Humber Lowlands is based on Smith (1958a. Brigg) and Godwin and Willis (1961 Immingham): from the Lincolnshire Marshes, on Swinnerton (1931) and Godwin and Willis (1964): from the Fenlands, on Godwin (1940a), Smith (1970) and Churchill (1970).

that at some time during Flandrian Ic, the Tees estuary was inundated at an altitude a little above -7.9 m O.D. On the basis of age, a correlation with Lytham I is suggested, but the altitude of the transgression in the Tees is high in comparison to the altitude of Lytham I in the type area. At various places in the estuary the upper part of the grey silty clay formation has a characteristic estuarine molluscan fauna of *Scrobicularia plana* da Costa, *Cerastoderma edule* Linné and *Littorina littorea* (Linné). Locally, at the mouth of the Billingham Beck, where it meets the River Tees, Agar (op. cit.) shows two peat beds intercalating the estuarine silts and clays, but these have not been found subsequently.

A detailed stratigraphic analysis of the intertidal peat beds of Hartlepool Bay has been made (Tooley in the press). A sample for radiocarbon dating from gyttja immediately below marine silt with *Scrobicularia plana* da Costa at an altitude of -2.05 m O.D. gave a date of 5285±120 (Hv. 4712), and a further sample from *Phragmites* peat at an altitude of -0.36 m O.D. immediately above the marine deposit gave a date of 5240±70 (Hv. 3459). The near synchroneity of the two radiocarbon dates from strata 171 cm apart can be interpreted as a result either of contamination by younger material of the lower sample or of rapid sedimentation after the onset of a late Flandrian II transgression. The date from *Phragmites* peat is from a level 10 cm below the *Ulmus* decline and is probably correct: the date from the gyttja cannot be corroborated independently in this way. None the less, the stratigraphy demonstrates unequivocally the presence of a single marine transgression along the western margin of Hartlepool Bay, which began towards the end of Flandrian II and ended shortly before the *Ulmus* decline. This transgression is registered during Lytham VI in north-west England, and undoubtedly represents the expression of this transgression in the Tees estuary.

There was no opportunity for marine and estuarine sedimentation during the Flandrian stage along the Yorkshire coast, until the Humber estuary is reached.

The lowlands and valleys adjacent to the Humber estuary in Lincolnshire and the East Riding of Yorkshire, and the coastal part of the Lincolnshire Marshes lack the systematic stratigraphic surveys and series of radiocarbon dates on the marine transgression limits necessary for detailed correlations with the patterns established in north-west England. Nevertheless, there is sufficient published evidence to single out the Humber Lowlands and the Lincolnshire coast as an area fundamental to the study of Flandrian marine sequences from the beginning of Flandrian II until the fourth millennium of Flandrian III. The following description is derived largely from the detailed stratigraphic and pollen analyses of Smith (1958a) in the valleys on the south side of the Humber estuary, from E. V. Wright and Churchill (1965) at Ferriby, from Gaunt and Tooley (1974) for the Humber estuary, and from Swinnerton (1931) at Chapel Point. In addition, relevant radiocarbon dates have been published by Godwin and Willis (1960, 1961, 1962, 1964) and Godwin and Switsur (1966).

At the Henderson Graving Dock, Immingham, a date of 6681±130 years BP (Q.401) on *Alnus* wood at -9.14 metres O.D. establishes a minimum age for the end of one transgression and a maximum age for the beginning of an early Flandrian II transgression at a recorded altitude of -9.09 m O.D. The ten centimetres thick peat bed is both underlaid and overlaid by a brackish water clay with *Phragmites* (Godwin and Willis 1961, p. 70). At the Market Place, Kingston upon Hull, Gaunt and Tooley (op. cit.) have recorded an estuarine sequence immediately above peat at an altitude of -9.11 m O.D., which occurred at some time after 6890±100 (IGS-C^{14}/100). The mean value for the beginning of this marine transgression in the Humber is 6785 radiocarbon years ago, which is rather close to the beginning of Lytham IV in north-west England.

On the east side of Spurn (TA 4235 1385), the culminating stages of this transgression have been dated more precisely to 6170±180 (Hv. 3359) at an altitude of -2.35 m O.D., although there is some evidence (Bisat 1952) that a *Scrobicularia* clay overlying peat at an altitude of -4.57 m O.D. in Hull Docks may record the beginning of a transgression, of which the data for Spurn record the culminating stages.

There is no incontrovertible evidence at present in the Humber Lowlands for a late Flandrian II transgression homologous to Lytham V, which is registered further north in Hartlepool Bay, but Bisat (op. cit.) has recorded a peat overlaid by a marine clay with *Scrobicularia plana* da Costa and *Cerastoderma edule* Linné at Kilnsea and at a higher altitude (c. 0.0 m O.D.) which may represent this transgression.

The Humber Lowlands and the Lincolnshire coast appear to have been affected most extensively by Flandrian marine transgressions from the early part of Flandrian III — a pattern similar to that recorded in the Fenlands.

In the Ancholme valley, Smith (1958a) has recorded the stages of marine inundation referable to late Flandrian II and Flandrian III. At the mouth of the Ancholme valley, the transgressive contact of a marine transgression is registered at an altitude of -3.51 metres O.D. The whole Ancholme valley was probably inundated by the sea for at Island Carr, Brigg, Smith (op.cit.) records a grey silty estuarine clay down to an altitude of about -6.5 metres O.D. There appears to be no interruption in the deposition of this clay throughout the valley, although its southern penetration beyond Redbourne Hayes is not known. Biogenic sedimentation in depressions on benches flanking the river valley ended as estuarine sedimentation continued. A situation in which this occurred is described by Smith (op. cit.) from Island Carr: here a transgression is recorded by a silty clay overlying brushwood peat from -2.82 metres to +1.00 metre O.D. The altitude of the transgressive surface rises westwards from Island Carr to +0.5 metres O.D. On the basis of the pollen spectra from the brushwood peat and on the provenance of the archaeological finds, Smith assigns the onset of estuarine conditions to the transition from the Bronze age to the Iron age, that is about 2500 years BP. This conclusion is now corroborated by a date on wood at Brigg, where fen brushwood gave way to alluvial clay at 2552±120 (Q.77).

It is difficult to envisage a marine transgression under way in the Ancholme valley from the early part of Flandrian II until the middle of the second millennium of Flandrian III, and continuing uninterrupted for a further unspecified period in Flandrian III, when at Spurn a marine regression is registered at 6170 years BP and, in north-west England, no less than four distinct marine transgressions are recorded at progressively higher altitudes above Ordnance Datum. It is possible that in the Humber Lowlands downwarping nullified the effect of slight falls of sea-level, and estuarine sedimentation was maintained within the valley at progressively higher levels. Indeed, estuarine sedimentation in the Ancholme valley appears to have been extremely rapid in the latter part of the third millennium of Flandrian III: an assay on wood from a prehistoric boat at Brigg, at an altitude of about +0.34 metres O.D. embedded within grey silty clay gave a date of 2784±100 (Q.78) and a recent assay on the Brigg raft resting on bluish, grey clay, at a similar altitude has yielded a date of 2543±100 (McGrail and Switsur 1975).

Estuarine sedimentation in the middle Ancholme valley continued up to about Ordnance Datum at Redbourne Hayes and over +1.0 metre O.D. at Island Carr, Brigg. The pollen assemblage from the remnant of upper peat at the former site does not permit any conclusion about a date for the culminating stage of this transgression in the Ancholme valley. The radiocarbon evidence from the Brigg raft indicates a post-2500 BP date for its culmination.

On the north side of the Humber estuary at North Ferriby, associated with the Ferriby prehistoric boats (McGrail and Switsur 1975), there is evidence for a retrogressive or still-stand stage in the transgression, registered in the Ancholme valley between about 6400 and 2500 years BP. E. V. and C. W. Wright (1947) show biogenic deposits up to -0.97 metres O.D. overlaid by grey marine clay.

There is a date on *Alnus* wood beneath prehistoric boat 3 embedded within estuarine, saltmarsh clay of 3120±105 (Q.715). The age of the onset of marine conditions at North Ferriby at an altitude of -0.97 metres O.D. (Wright and Churchill 1965) clearly pre-dates the age of the wood by an unknown period.

Further south, in Lincolnshire, the most complete coastal sequence derives from the detailed descriptions of Swinnerton (1931) north of Chapel Point. In addition, H. and M. E. Godwin (1934) have carried out pollen analyses of the upper and lower peat beds, and Smith (1958a) has also carried out further analyses on the upper peat bed. Furthermore, there are six radiocarbon dates on material from these coastal deposits.

Swinnerton does not give a stratigraphic description of the deposits, and the following has been derived from his diagrammatic section from which altitudes have been scaled:

Stratum	Height O.D. metres	Description
6	+2.31 to +0.24	Clay, well-laminated, soft and sloppy, Purple tint. Towards the base, valves of *Scrobicularia plana*, and *Cardium edule* in their living postures.
5	+0.24 to 0.00	Upper peat. Locally rich in *Phragmites*. Peat well-stratified. Towards the top of Stratum 5, becomes a woody detrital peat with *Salix* and *Taxus*.
4	0.00 to -0.36	*Phragmites* clay. Transition between Strata 4 and 5.
3	-0.36 to -1.64	Clay with *Triglochin, Phragmites, Juncus, Limonium* and *Armeria*.
2	-1.64 to -2.19	Lower peat. A compacted, woody detrital peat, with *Alnus, Betula* and *Quercus*. Sharp contact between Strata 2 and 3.
1	-2.19	Boulder clay, where exposed, bluish and podsolised.

A sample of peat from the top of Stratum 2 has been dated to 3943±100 years BP (Q.685) at a given altitude of −1.83 metres O.D., which is well within the altitudinal limits given here for Stratum 2. This date is maximal for the onset of marine conditions here, for Swinnerton notes that erosion of the lower peat took place before the deposition of Stratum 3. That a transgression, recorded by Stratum 3 at Chapel Point, was under way elsewhere in north-east England is confirmed by radiocarbon assays on intertidal molluscs from Stoneferry A, north of the Humber estuary (Gaunt and Tooley 1974), which yielded dates of 3775±100 (IGS-C^{14} 98I) on *Cerastoderma edule* Linné at an altitude of −3.80 m O.D., and 3435±200 (IGS-C^{14}98II) on *Macoma balthica* (Linné) at a mean altitude of −3.40 m O.D. The difference in altitude between the Stoneferry samples and the Chapel Point sample is probably explicable in terms of the provenance of the dated material: at Stoneferry, the molluscan samples record an altitude close to the contemporary low-tide level, whereas at Chapel Point the evidence points to a high-tide level.

Strata 3 and 4 indicate changing environmental conditions at Chapel Point during the transgression, the culminating stages of which are represented by Stratum 4. A radiocarbon assay on peat from the base of Stratum 5 at +0.15 metre O.D. gave a date of 3340±110 (Q.686). An earlier assay on wood from the top of Stratum 5 provided a date of 2455±110 (Q.81). Although there is no means of establishing the exact relationship of this dated material to the second marine transgression registered at Chapel Point as Stratum 6, two internally consistent dates on valves of *Scrobicularia plana* da Costa from the base of Stratum 6, where it overlies Stratum 5, on an eroded surface, both gave ages of 2630±110 (Q.678, Q.688). Where Stratum 5 is overlaid directly by clay, a sample of *Phragmites* peat from the top of the stratum at about +0.15 m O.D. gave a date of 2815±100 (Q.844).

The well-laminated estuarine clay recorded as Stratum 6 at Chapel Point was undoubtedly laid down contemporaneously with the stiff, light brown silty clays recorded at Island Carr, which are unequivocally of marine origin (Wright and Churchill 1965). At Chapel Point, the transgression occurred at an altitude of about +0.15 m O.D., about 2815 years BP, while at Brigg it occurred at an altitude of +0.34 m O.D. about 2784 years BP.

The data from the Humber Lowlands can be summarised and correlated tentatively with those from north-west England: the regressive contact of an early Flandrian transgression is registered at Immingham, and correlates with a late Flandrian I/early Flandrian II transgression in north-west England, specified as Lytham III. A second transgression is recorded at Immingham and the Market Place, Kingston upon Hull, a little after 6681 years BP, and the culminating phase is probably registered at Spurn at about 6170 years BP: a transgression with these time limits falls well within the age limits of Lytham IV established in north-west England. There is no evidence for a mid-Flandrian II transgression in north-east England, but at Hartlepool, a late Flandrian II transgression of short duration (5285 to 5240 years BP) occurred during the early part of Lytham VI. A fourth transgression is recorded at Chapel Point, where absolute time limits for the transgression have been established at 3943 to 3340 years BP and supported by data from Stoneferry A, which can be correlated with Lytham VII. The transgressive contact of a fifth transgression is recorded at Chapel Point at 2815 and at Brigg at 2784, although no evidence is available for the exact date of the removal of marine conditions; this transgression can be correlated with Lytham VIII (Fig. 39).

There is stratigraphic and archaeological evidence at South Ferriby for a sixth transgression, together with evidence for the culminating stages of the fifth transgression recognised here. At South Ferriby, Smith (op. cit.) describes a light brown silty clay overlaid by dark brown clayey peat and in turn by a light brown silty clay. The upper surface of the lower light brown silty clay undulates with height limits of about +1.4 to +2.0 metres O.D., and in the upper 15 cm of this stratum Romano-British pottery has been recovered. This stratum may represent the culminating phases of the fifth transgression, while the upper light brown silty clay may be a sixth transgression of medieval age.

The foregoing scheme represents a synthesis of the available data from the Tees Lowlands, the Humber Lowlands and the Lincolnshire coast. The data derive from scattered sites, and any pattern thus derived must be regarded as tentative until detailed stratigraphic and microfossil analyses establish the spatial pattern of these Flandrian marine deposits, and a series of radiocarbon dates fixes the ages of the transgression sequences.

The Fenlands

The stratigraphy of the perimarine, lagoonal and tidal flat zones of the Fenlands and adjacent areas has been subject to detailed examination by Skertchley (1877), by Godwin from 1933 onwards (H. and M. E. Godwin 1933a, Godwin and Clifford 1938, Godwin 1940a) and by Jennings (1952). Godwin's work largely covered the area south of a line from Peterborough through Wisbech to King's Lynn, from the meres in the perimarine zone along the western fen margin through lagoonal and tidal flat deposits to the present coast. The area to the north around Spalding has been subject to some examination by Jennings, the results of which have been published recently by Smith (1970). Both Smith (1970) and Churchill (1970) have attempted a correlation within the Fenland and adjacent areas, but both stress the inadequacy of the available data for the task. It is a cause of some regret that radiocarbon dating methods were not available when Godwin undertook his stratigraphic and pollen analytical work in the Fenlands; subsequent attempts to establish a chronology of Flandrian marine successions in the Fenland (Willis 1961: Churchill 1970) have suffered from the dearth of consistent radiocarbon dates on the marine transgressions recognised over a wide area. In both cases, dates from archaeological sites had to be used to place minimum or maximum ages on the beginning or end of a transgression. Furthermore, a considerable body of stratigraphic data derives from engineering borehole records, which do not allow precision in the establishment of marine transgressions based on altitudes alone, and some of the difficulty in integrating the available data from the Fenland must derive from this fact.

Notwithstanding these difficulties, the considerable volume of data from the Fenland indicates a pattern of Flandrian marine transgression sequences similar to those recorded both in the Netherlands and in north-west England. The age of the earliest transgression recorded at Guyhirn, King's Lynn and around Spalding has not been established, but the present Fenland was probably transgressed somewhat later than elsewhere in England and the southern North Sea Basin; the present Lancashire coast had been transgressed by the middle of Flandrian I (at Lytham, a little after 8390±105, and in Morecambe Bay from 8925±200 BP onwards) and the Dutch coast towards the end of Flandrian I (at Uitgeest, an assay on the top of the lower peat gave a date of 7540±70 BP and at Velsen 7140±70 BP (Jelgersma 1961)).

During Flandrian I the present Fenlands must have been well above sea-level, for there is no evidence that it had been affected by the rise of sea-level. The area between the land limit of the Fen Clay and the western limit of the peat probably comprised the perimarine zone of the Fenland and appears to have been initiated no earlier than 4700 years BP (Q.544 Wood Fen, Ely, 4195±110, and Q.545 Woodwalton Fen 3415±110, Q.474 Glass Moor

4345±110 and Q.130 Adventurer's Fen Wicken 4605±110). The elevation of the freshwater surface here is intimately related to the deposition of the Fen Clay further east.

The Brancaster peats at Judy Hard on the Norfolk coast described by H. and M. E. Godwin (1934) are undoubtedly of late Flandrian I age, and are recorded at altitudes of from −3.00 to −0.4 metres O.D., but the relationship to their contemporary sea-level has not been established. The Golf Club Foreshore peats that the Godwins describe may be related to sea-level: the lower peat intercalates a stiff blue clay probably of estuarine origin, for which there are altitudinal limits of −1.67 on the regressive contact and −0.76 metres O.D. on the transgressive contact. The peat can probably be assigned to Flandrian II.

Evidence for Flandrian marine successions during Flandrian I comes from the North Sea, off the East Anglian coast.

Erdtman had carried out pollen analyses on moorlogs dredged from the Dogger Bank as early as 1926, and H. and M. E. Godwin (1933b, 1934) published four pollen spectra from moorlogs dredged from the Leman and Ower Banks. In one sample there was a preponderance of *Pinus* and *Corylus* over other taxa, and in another of *Betula* and *Pinus*. On this basis, the peat was assigned to the Boreal (Flandrian I); greater chronological precision was obtained from a radiocarbon date of 8422±170 (Q.105). There is no evidence from the pollen assemblage of approaching marine conditions, and this date is probably maximal. Jelgersma (1961) obtained five samples on a line north of Lowestoft-Ijmuiden and carried out pollen analyses on the dredged samples: approaching marine conditions are probably registered in three of the samples on the basis of low Chenopodiaceae frequences. The ages of the samples fall within the limits 8700 to 8400 radiocarbon years ago. Samples from the North Sea basin related to contemporary sea-levels during the Flandrian stage, have not yet been obtained. Nevertheless both Godwin (1956) and Jelgersma (1961) have indicated stages in the closure of the North Sea land-bridge: Godwin has shown that closure probably occurred during the Boreal, while Jelgersma indicates that closure had taken place by about 8600 years BP.

Kolp (1976) has identified a series of terraces at −60 m, −45 m, −30 m, −24 m, −19 m, −13 m and −7 m N.N. in the Southern Baltic Sea and the North Sea associated with stages in the restoration of sea-level during the Flandrian stage. The −45 m terrace south of the Dogger Bank comprises a freshwater basal peat of pre-Boreal age, overlaid by brackish water sands, silts and gravels with lenses of peaty gyttja of young Boreal age, about 7800 years BP. If the lenses of peaty gyttja are *in situ*, then it does indicate that more of the North Sea land-bridge remained intact at the opening of the Boreal than had hitherto been expected at this time.

There is no evidence for the age of the first marine transgression recorded in the Fenlands. Between Guyhirn Railway Bridge and Cross Guns, Godwin and Clifford (1938, Fig. 27, p. 370) show a marine transgression between about -10.5 and -9.5 metres O.D., rising south-westwards to about -8.0 metres O.D. attenuating and intercalated by a lower peat. East of Spalding, Smith (1970) has recorded the culminating phase of a marine transgression at -7.9 metres O.D. At St. German's, Godwin has recorded a transgressive phase at an altitude of -7.11 metres O.D. At this site the following stratigraphic succession has been recorded:

Stratum	Height O.D. metres	Depth cm.	Description
	+5.48 to +1.83	000 to 365	Made Ground
J	+1.83 to +1.23	365 to 425	Blue Clay
H	+1.23 to +0.93	425 to 455	One foot peat bed
G	+0.93 to +0.02	455 to 546	Brown, silty clay, passing into F
F	+0.02 to -0.89	546 to 637	Blue clay with *Scrobicularia*
E	-0.89 to -1.49	637 to 697	Two foot peat bed
D	-1.49 to -5.14	697 to 1062	Blue clay, mottled brown in places with *Cardium edule*
C	-5.14 to -5.29	1062 to 1077	Six inch peat bed with oak tree stool
B	-5.29 to -7.11	1077 to 1259	Blue, buttery clay. Locally a gravelly bed, with grey and brown sandstone pebbles, replacing the top foot of clay.
A	-7.11 to -7.13	1259 to 1261	One inch peat bed
	-7.13	1261+	Kimmeridge Clay

Godwin has published a pollen diagram from Strata A, C, E and H (H. and and M. E. Godwin 1933a), and an assay on oak wood from Stratum C gave a date of 4690±120 years BP (Q.31). This gives a minimum date for the culminating stage of the transgression recorded in Stratum B and a maximum date for the onset of marine conditions registered in Stratum D.

On the basis of a single pollen spectrum from Stratum A, in which *Alnus* and *Tilia* are present, Godwin assigns the lowest level to the post-Boreal. It is probable that the transgression recorded in Stratum B occurred towards the end of Flandrian II, especially as its culminating stage is registered at the beginning of Flandrian III. If this is the case, then this transgression can be correlated with Lytham VI, for which there are time limits of 5570 to 4897 years BP, and with the transgression recorded along the western margins of Hartlepool Bay.

The deposition of Fen Clay, characterised by Stratum D at St. German's, began a little after 4690±120, and penetrated as far as Glass Moor, Ramsey, where cones of *Pinus sylvestris*, in a wood peat overlaid by Fen Clay, gave a date of 4345±120. Within the perimarine zone, unaffected by sedimentation of Fen Clay directly, but registering a rise in the freshwater table, dates of 4195±110 (Q.544) and 4605±110 (Q.130) have been obtained from Wood Fen, Ely, and Adventurer's Fen, Wicken, respectively. The culminating phases of the Fen Clay transgression have been recorded from Saddle Bow and Denver Sluice. At Saddle Bow the culminating phase occurred at 3915±120 (Q.490) and at Denver Sluice a little before 4085±110 (Q.264). In the Witham valley, south-west of Woodall Spa, Valentine and Dalrymple (1975) have recorded the end of peat sedimentation and the beginning of Fen Clay deposition at 4162±130 (HA.150) and 3945±100 (IGS-C^{14}/109). No transgression with these time limits has been proved unequivocally in north-west England, but in the Netherlands, the last Calais transgression, C-IV, has time limits of 4550 to 3750 BP (Hageman 1969), and the deposition of the Fen Clay in the Fenland and Calais IV deposits in the Netherlands appears to be approximately synchronous.

A period of about two thousand years, during which the upper peat accumulated, came to an end when a transgression brought marine to estuarine sedimentation landward of Denver Sluice. A typical section through the silt into the underlying upper peat is given by Forbes *et al.* (1958) from Saddle Bow (see p. 161) and an approximate ground height of +1.5 metres O.D. has been scaled from data given by Churchill (1970, Fig. 7).

Two assays on material from an adjacent site, taken from the upper 4 cm of a stratum equivalent to Stratum 3 described here, gave two consistent dates of 1875±110 (Q.549) 0 to 2 cm from the contact between Strata 3 and 4, and 2070±110 (Q.550) 2 to 4 cm below the same contact. On this basis, Godwin argued (Godwin and Willis 1961) that the grey silt constituting Stratum 4 was laid down during a transgression that began in Romano-British times. An assay on fen peat subjacent to estuarine silt at Magdalen Bend gave a date of

Stratum	Height O.D. metres	Depth cm.	Description
4	+1.50 to +0.59	000 to 091	Grey silt, weathering brown above, penetrated from the surface by *Phragmites* rhizomes. The lower 10 cm. contained numerous seeds of *Triglochin* and *Salicornia* (Churchill 1970).
3	+0.59 to -0.17	091 to 167	Chocolate-brown, fen peat with wood and *Phragmites* rhizomes. The upper 4 cm. contained charcoal and the seeds of *Sparganium*, *Hydrocotyle* and *Eupatorium* (Churchill 1970).
2	-0.17 to -0.29	167 to 179	Brown clay penetrated by *Phragmites* rhizomes. Transition to Stratum 1.

3305±120 (Q.547) and Godwin suggests that here the upper surface of the peat has been eroded, in which case the date does not record the onset of marine conditions. However, the altitude at which the transgressive contact occurs, estimated from Churchill's Figure 7 (1970) to -1.2 metres O.D., clearly places this transgression within a different height class, and more nearly within the culminating stages of the transgression that deposited the Fen Clay, represented by the stratigraphy from Saddle Bow by Stratum 1.

Churchill (1970) re-examined the monolith from Saddle Bow from which samples were taken for radiometric analysis, and three further samples were re-submitted from Stratum 3: these gave dates of 2495±110 (Q.805) from the top of Stratum 3, 2275±100 (Q.806) from 12 cm below the 3/4 boundary and 2377±100 (Q.807) from 30 cm below the 3/4 boundary.

This ambiguous evidence from the floodplain of the River Ouse certainly does not permit precision in establishing the age of a transgression that was undoubtedly under way before the Roman occupation, and probably continued until the second century A.D. (Churchill 1970).

A final marine transgression is inferred from flooding of the southern Fenland margins and alluviation of the fenland drainage channels up to +2.0 metres O.D. during a period ascribed to the third century A.D. by Churchill on the basis of archaeological remains. The creation of roddons, described by Godwin (1940a) at this time, is analogous to the Gorkum and Tiel deposits of river channels and flood plains of the Netherlands, described by Hageman (1969) from the perimarine zone, and causally related to the movements of sea-level registered by marine sedimentation further seaward. The alluviation that Churchill describes from the Fenlands in the third century A.D. can be correlated with a marine transgression registered in the Netherlands between A.D. 250 and 600 (1700 to 1300 BP) as Dunkirk II, which generated the sandy and silty clays of Tiel II in the river valleys of the perimarine zone.

A tentative correlation can be made with episodes in north-west England,

and these are summarised in Fig. 39.

In 1976 (Tooley unpublished) stratigraphic and pollen analytical investigations west of Guyhirn on Adventurer's Land (TF 35670182) have
indicated the existence of five marine transgressions in the valley of the River
Nene. At the first sampling site (AL-2), the transgressions have the following
altitudinal limits:

Transgression 1.	Strata 6-7	-7.00 to -5.83 m O.D.
Transgression 2.	Strata 9-10	-5.66 to -4.51 m O.D.
Transgression 3.	Stratum 12	-4.36 to -3.66 m O.D.
Transgression 4.	Stratum 14	-3.59 to -3.22 m O.D.
Transgression 5.	Stratum 19	-2.96 to -1.62 m O.D.

Transgression 1 overlaid a gyttja with monocotyledonous roots that gave way
in turn to a woody detrital peat and a sandy silt at -7.74 m O.D., before gravel
was reached at -7.80 m O.D. The gyttja yielded a tree pollen spectrum of
3% ΣAP *Betula*, 2.5% *Pinus*, 6% *Ulmus*, 43% *Quercus*, 16% *Tilia* and 29%
Alnus, which points to a Flandrian II age: the basal biogenic deposit is no
older than 7000 years BP, and all the transgressions recognised on
Adventurer's Land are younger than this. On the basis of altitude,
Transgression 1 appears to be registered at St. German's as Stratum B, but
during the deposition of Stratum D at St. German's, Transgressions 2-5 were
recorded on Adventurer's Land. Further correlations must await radiocarbon
assays from the intercalated biogenic strata.

The difference between marine successions in the Fenlands and north-west
England is the late arrival of marine conditions, so that at least four marine
transgressions had been registered in north-west England before the Fenlands
were affected extensively. Once marine transgressions began in the Fenlands,
they affected extensive areas, and whereas in north-west England they became
progressively confined to embayments adjacent to estuaries during Flandrian
III, there is no evidence of a similar confinement in the Fenlands. A series of
dated sections from the perimarine, lagoonal and tidal flat zone of the
Fenlands is needed before the age and extent of marine transgression
sequences can be established satisfactorily for this area; such sites have been
proved by Godwin at Guyhirn and King's Lynn, and both merit detailed re-
examination.

The Somerset Levels

Flanking the Bristol Channel and the Severn Estuary are low-lying areas
containing biogenic sediments of Flandrian age. These sediments are
interrupted by clays, silts and sands of estuarine origin, and together provide
evidence of relative changes in sea-level during the Flandrian stage, and the
extent and age of transgression sequences, which are the expression of these
relative changes in sea level.

Areas that contain evidence for marine transgression episodes in this region include the tidal flat, lagoonal and perimarine zone of Barnstaple Bay (Churchill 1965a; Kidson 1971), Bridgwater Bay (Kidson 1971; Kidson and Heyworth 1973), the Somerset Levels (Godwin 1941, 1943, 1948, 1955 and 1960; Clapham and Godwin 1948; Hawkins 1971; Kidson 1977), the Yeo and Kenn valley (Gilbertson unpublished), the Gordano valley (Jefferies *et al.* 1968), the Vales of Berkeley and Gwent, particularly the Wentlooge and Caldicott Levels, and Swansea Bay (von Post 1933: Godwin 1940b, 1943). Of these, the papers published by Godwin contain the most detailed stratigraphic and pollen analytical details, which permit, with the radiocarbon dates obtained subsequently, a tentative correlation with the sequences established in north-west England.

Two sites at Shapwick Station (Godwin 1941) and at East Brent (Hawkins 1971) are critical for the establishment of a Flandrian marine chronology in the Somerset Levels, but there does not appear to be either a single site or a restricted area within the Levels where a complete record of transgression sequences has been registered. On the basis of published data from sites throughout the Somerset Levels and in adjacent areas, it is possible to recognise six distinct marine transgressions in this area.

Evidence for the first approach of marine conditions during the Flandrian stage comes from Baglan Burrows near Port Talbot: here, Godwin and Willis (1964) have described a compressed, freshwater peat at −18.89 metres O.D. overlaid by marine sands and silty clays. An assay (Q.663) on a thin slice of this material gave a date of 8970±160. There can be little doubt that this peat was accumulating within the perimarine zone of Swansea Bay, and that biogenic sedimentation was terminated by estuarine sedimentation at the site. There is only indirect evidence, in the elevation of the freshwater table at the site for approaching marine conditions, and both the date and the height of the isolation contact must be regarded as maximal. At Dunball in the Somerset Levels, Hawkins (1971) has recorded a peat at −20.7 metres O.D. dated at 8360±145 BP (I-4315), and at Kingston Seymour in the River Yeo valley, at −4.6 metres O.D. a basal peat has been dated 8690 years BP. It is possible that this transgression is also registered at East Brent at −20.16 metres O.D. and at Shapwick Station at −13.15 metres O.D. for its transgressive contact. The removal of marine conditions is registered at an altitude of −11.17 metres O.D. at Shapwick Station, and at East Brent at a height of −18.48 metres O.D. On the basis of the considerable altitudinal differences of the transgression and regression contact at East Brent and Shapwick Station, the succession at East Brent could be reinterpreted as an earlier marine episode, antedating the first transgression recognised at Shapwick Station.

There can be little doubt of the synchroneity of this transgression in the Bristol Channel and Somerset Levels with Lytham I, for which there are time limits of 9270 to 8575 years BP. The transgressive phase of Lytham I is registered in Morecambe Bay at 9270 years BP, at an altitude of −17.60 metres O.D., and the regressive phase at Heyhouses Lane, Lytham, at 8575 years BP, at an altitude of -9.75 metres O.D. The time and altitude of this transgression in two areas in western England, some 280 km apart, are remarkable.

A second transgression can be identified and seems to have affected extensive areas of the Somerset Levels. At Shapwick Station, the transgressive contact is registered at -11.07 metres O.D. where the woody, detrital peat is overlaid by a sandy clay. In Bridgwater Bay, Kidson (in Buckley and Willis 1969) has recorded the transgressive stages of this transgression between -9.14 and -6.09 metres O.D. during the culminating stages of Flandrian I, and Hawkins (in Welin, Engstrad and Vaczy 1972) has recorded the same transgression at East Brent. There appears to be no evidence for the culminating stages of this transgression, which falls within the time limits of Lytham III established at 7605 to about 7200 years BP. In north-west England, Lytham III occurs at an altitude of -2.40 metres O.D., considerably higher than in the Somerset Levels. Estuarine sedimentation appears to have been maintained for over a thousand years in the Somerset Levels, at the end of which time an extensive removal of marine conditions occurred. The regressive contact of this transgression is recorded at Witchey Bridge at about -4.6 metres O.D. by Godwin (1943) and at Stolford at -2.7 metres O.D. by Kidson at some stage before 6230±95 (NPL. 148) (in Callow and Hassall 1968). Outside the Somerset Levels, the same regressive contact has been registered by Godwin and Willis (1961) at Margam, where material from the base of a peat bed overlying brackish water clays at -3.2 metres O.D. gave a date of 6184±144 (Q.275). The culminating stage of this transgression is registered in north-west England as Lytham IV, for which there is a mean date of 6157 years BP.

A third transgression is recognised in the Somerset Levels and is correlated with Lytham VI, for which there are time limits of 5570 to 4897 years BP. This transgression constitutes a late Flandrian II/early Flandrian III transgression in north-west England, and its culminating stages witnessed an extensive removal of marine conditions. A similar pattern occurs in the Somerset Levels, though the inception occurs later, and the culminating stages earlier, than in Lancashire.

At Witchey Bridge, the inception of marine sedimentation is recorded at a height of about -4.0 metres O.D. (Godwin 1943). At Stolford, Kidson (in Callow and Hassall 1968) records a date of 5380±95 on peat at -0.52 metres O.D., which is a maximum date for the beginning of marine sedimentation here. The culminating stages of the transgression are recorded extensively, at a time pre-dating the *Ulmus*-decline on the Levels. In 1943, Godwin showed a more or less complete evacuation of estuarine conditions from the River Brue valley, from Decoy Pool to Puriton Drove, at an altitude close to Ordnance Datum. At Shapwick Heath, the *Ulmus*-decline is recorded 120 cm above the regressive contact of this transgression (Godwin 1941) and 56 cm above the contact at Site F on Shapwick Heath, recorded by Coles *et al.* (1973). Subsequently, a series of radiocarbon dates has corroborated this age, and placed the culminating stages of the transgression towards the end of

Flandrian II. At Shapwick Heath, an assay on clayey *Phragmites* peat overlying estuarine clay at an approximate height of +1.8 metres O.D. gave a date of 5510±120 (Q.423, Godwin and Willis 1961). On Tealham Moor, two dates on the same isolation contact were consistent with the date from Shapwick Heath, 5412±130 (Q.120) and 5620±120 (Q.126) (Godwin and Willis 1959, 1961). Godwin (1955) also gives height data for the isolation contact of this transgression: at Glastonbury Lake village (GLV-I) it is +0.55, at Drake's Drove +0.53 and at Godney Moor (GO-IX) +0.01 metres O.D. At Combwich (Godwin 1941), it is registered at +1.62 metres O.D., and in the Gordano valley (Jefferies *et al.* 1968) at heights between +1.21 and +3.04 metres O.D.

There is evidence for a fourth transgression that has been proved early in Flandrian III from Kenn, Avonmouth and Portbury (Hawkins 1971: Welin, Engstrad and Vaczy 1972). In the Fenlands also, the Fen Clay has time limits of 4690 to 3915 years BP, and this transgression has been correlated with Calais-IV in the Netherlands, but there is no unequivocal evidence for this transgression in north-west England.

The fifth transgression to affect the Somerset Levels has been recorded at Kingston Seymour by Hawkins (1971). Here, the inception of marine conditions is registered at a height of +1.3 metres O.D., at 3690±110 (I.4846). Kidson has recorded the transgressive phase of the same transgression at Stolford, where, at a height of +1.0 metres O.D., peat subjacent to a marsh clay has been dated at 3460±90 years BP (NPL.146). The regressive contact of this transgression is recorded by Hawkins at Kingston Seymour at +4.1 metres O.D. 3350 years BP; the altitude appears somewhat high for the culminating stage of a transgression of this age. This transgression is homologous to Lytham VII with time limits of 3700 to 3150 years BP.

The sixth and final transgression recorded in the Somerset Levels was regarded as Romano-British (Godwin 1955) from the included Roman remains. The transgressive contact of this transgression occurs at a height of +3.29 metres O.D. at Glastonbury Lake Village (GLV-II) where it overlies a dark brown detritus mud with *Menyanthes* fruits, at +2.26 metres O.D. at Godney Moor (GO-IX) and at +2.28 metres O.D. at Drake's Drove. There is no radiocarbon date to establish the age of the onset of this transgression in the Somerset Levels, but the approaching marine conditions may have caused the elevation of the freshwater table in the perimarine zone of Shapwick Heath, and Godwin has noted (Godwin and Willis 1960; Godwin 1966) that the response to the waterlogging of the bog surfaces led to the construction of trackways between 2470 and 2850 years BP. At Llanwern, on the north side of the Severn Estuary, Godwin and Willis (1964) have dated peat immediately subjacent to a silty-marine clay at a height of +2.8 metres O.D. to 2660±110 (Q.691). The removal of marine conditions probably resulted in a fall of the freshwater table, in the perimarine zone, which can be inferred at Shapwick Heath by the replacement of *Cladium* peat representing the first flooding horizon by *Sphagnum* peat, containing *Betula*: an assay on the birch wood (Q.36) gave a date of 2250±110. In north-west England, Lytham VIII has time limits of 3090 to 2270 years BP. Murray and Hawkins (1976) argue against a Romano-British transgression in the Somerset Levels on the evidence that the foraminiferal assemblages from the clay indicated continuous sedimentation.

The foregoing description achieves no more than a collation of the available data from the Somerset Levels and adjacent areas, and suggests that the pattern of transgressions identified from these data has similarities with that of transgressions in north-west England.

1.2. NORTH-WEST EUROPE

Along the margins of the English Channel, North Sea and Baltic Sea (Figure 1), there are opportunities for quiet water sedimentation within the tidal flat, lagoonal and perimarine zones, in which, both directly and indirectly, marine transgressions have been registered during the Flandrian stage. Where the age of these transgressions has been established in the Netherlands and Sweden, for example, tentative correlations can be made with the sequences in north-west England (Fig. 40). The comparison could be extended to Denmark (Iversen 1937; Mikkelsen 1949; Krog 1965, and Krog and Tauber 1973), to Germany (Brand *et al.* 1965; Duphorn *et al.* 1973; Geyh 1969, 1971; Streif 1972) and to Finland (Eronen 1974), where similar patterns of marine transgressions during the Flandrian stage have been established.

The Netherlands

In the Netherlands, detailed stratigraphic, pollen and radiometric analyses by members of the Dutch Geological Survey, particularly by Jelgersma (1961), permit the establishment of the age and the extent of the transgressions recognised along the Dutch coast. Four lithologic units were recognised by Jelgersma in the coastal area: a lower peat of Boreal and early Atlantic age in which approaching marine conditions were registered at some sites; Old Tidal Flat deposits, equivalent to the *Assise de Calais* (Dubois 1924), comprising marine sands, silts and clays, and interrupted by thin peat layers. These deposits accumulated from the opening of the Atlantic until the early Sub-Boreal, and are overlaid by the Upper or Holland peat, which accumulated during the Sub-Boreal and well into the Sub-Atlantic locally. Accumulation of the Holland peat was interrupted by a *Cardium* transgression from about 3600 to 2660 years BP, and is overlaid by Young Tidal Flat deposits, equivalent to the *Assise de Dunkerque* (Dubois 1924). These deposits can be subdivided on the basis of the intercalated biogenic strata into three distinct transgressions.

The subdivision of the Flandrian marine sequences in the Netherlands has since been refined, and the following scheme has been described by Hageman

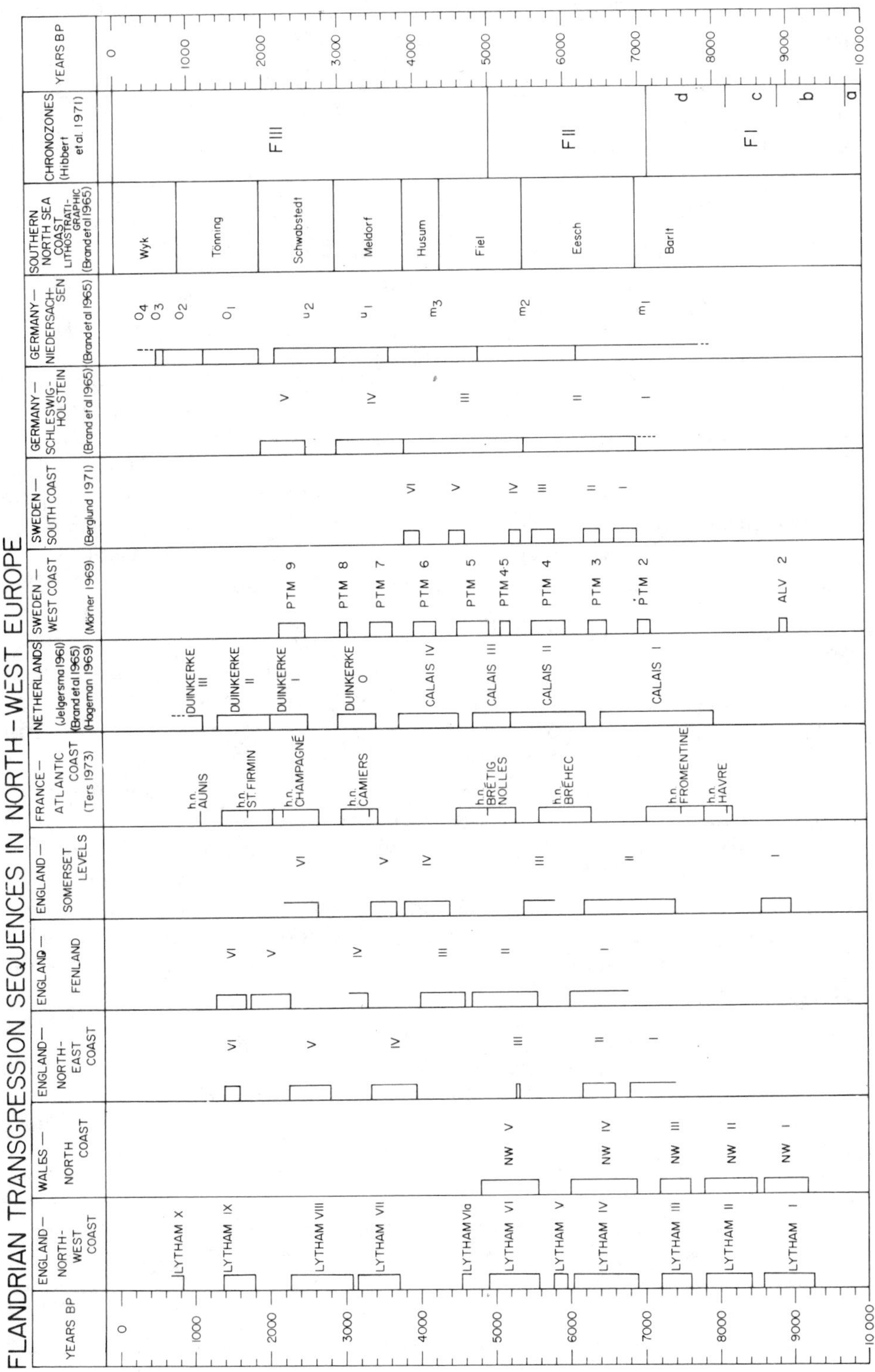

Fig. 40. Scheme of Flandrian Transgression Sequences in North-west Europe. The evidence from the north-west coast of England is summarised in Fig. 36, and from north-east England and the Fenlands in Fig. 37. The evidence from the Somerset Levels is based on Godwin (1941), Kidson (1971, 1977), and Hawkins (1971). The sources for the age of marine transgressions in the Netherlands, Sweden and Germany are shown on the diagram.

(1969) and Oele (1977). The period from about 8000 to 4000 years BP during which the Old Tidal Flat Deposits were accumulating, has been subdivided into four periods characterised by four distinct transgressions, separated by biogenic sediments:

	Hageman (1969)	Oele (1977)
Calais IV (C.IV)	4550 to 3750 years BP	4600 to 3800 years BP
Calais III (C.III)	5250 to 4750 years BP	5300 to 4600 years BP
Calais II (C.II)	6250 to 5250 years BP	6300 to 5300 years BP
Calais I (C.I)	7950 to 6450 years BP	7500 to 6300 years BP

The time limits on the transgressions derive from dates from the eastern margin of the tidal flat zone: further westward, the onset of each transgression occurred earlier than the limit given, and the culminating stages later. Indeed, Jelgersma (1961) makes the point that the regressive contact of the Old Tidal Flat Deposits (Calais IV) is younger near the coast than at the point of maximum penetration of these deposits. Ente *et al.* (1975) have proposed a subdivision of Calais IV based on lithostratigraphy and radiocarbon dating. Calais IV A1 is dated at 4700 to 4400; Calais IV A2 from 4400 to 4100 BP, and Calais IV B from 4100 to 3750 BP. The maximum thickness of marine deposits referable to the Calais period is 15 metres, occurring between height limits of approximately -4 to -18.5 metres N.A.P., which, for the latter figure, is well within the limits established by Dubois (1924) for the *Assise de Calais*.

From about 3500 years BP to the present a further four marine transgressions are recognised, and are characterised by thin clay beds and intercalated biogenic strata rarely thicker than two metres and of limited geographical extent. These transgression sequences are correlated with the *Assise de Dunkerque*, established by Dubois, on the basis of the palaeontology of the marine deposits which contain faunal assemblages similar to contemporary assemblages, such as *Cardium edule* Linné and *Scrobicularia plana* da Costa. The following four transgressions are recognised in the Netherlands after 3500 years BP to which the name *Dunkirk* has been given after the type area in Northern France:

Dunkirk III (D.III)	1150 to date years BP	1100 and later
Dunkirk II (D.II)	2050 to 1350 years BP	1700 to 1350 years BP
Dunkirk I (D.I)	2550 to 2050 years BP	2500 to 1950 years BP
Dunkirk 0 (D.0)	3450 to 2950 years BP	3450 to 2950 years BP

Hageman makes the important distinction between the areas affected directly and indirectly by a marine transgression, and this distinction has been applied to successions in north-west England. He distinguishes three areas in the northern and western margins of the Netherlands: a Lagoonal and Tidal Flat area, a Perimarine area and a Beach and Dune Area. The lagoonal and tidal flat area was affected directly by Calais and Dunkirk transgressions, and the sediments comprise clays, silts and sands of estuarine and marine origin, intercalated by biogenic sediments, comprising both saltmarsh, reedswamp, freshwater muds and terrestrial peats. Landward of this area is the perimarine zone, which has been affected indirectly by the marine transgressions registered in the lagoonal and tidal flat area. Hageman defines the perimarine zone as 'the area where sedimentation took place under the direct influence of relative sea-level movements, but where marine or brackish sediments themselves are absent . . . giving rise to the development of thick, fluviatile clay layers, alternating with peat layers, cut by fossil gulleys filled up with sand'. Sediments typical of the perimarine zone are defined in the Netherlands as Gorkum and Tiel deposits, which are genetically related to the Calais and Dunkirk transgressions registered in the lagoonal and tidal flat zone. An homologous pattern has been described by Godwin in the Fenlands, where river channels have been recorded in the Huntspill area. Biogenic sediments comprise limnic deposits, telmatic deposits, such as *Phragmites* and *Cladium* peat, and terrestrial deposits, such as woody and herbaceous detrital peat. The seaward margin of the perimarine zone dilates during marine regressions and contracts during transgressions: in north-west England, the freshwater lakes of Black and White Otter, Gettern and Downholland Mere in south-west Lancashire were initiated by marine transgressions that failed to reach their margins, but elicited a regional rise in the freshwater table, yet these meres overlie alternating clay and peat deposits characteristic of the lagoonal and tidal flat zone, registering marine transgressions Lytham II to VI.

The relationship between the age of marine transgressions in north-west England and the Netherlands is shown in Fig. 40. The lack of detailed synchroneity derives in part from the position within the respective lagoonal and tidal flat zones from which samples were taken for radiocarbon dating, for the time limits do not represent average time limits for each transgression, and in part from local conditions, which can advance or retard the registration of a marine transgression. Notwithstanding the detailed age differences, the over-all pattern of marine sequences in the Netherlands and north-west England is similar: there is no homologue of Lytham I in the Netherlands, which is to be expected, for the evidence of Lytham I and earlier transgressions resides in the northern North Sea Basin, the Celtic Sea and the approaches to the English Channel. In north-west England, there is so far no direct evidence for Calais IV. There is indirect evidence from the dune chronology that a transgression was under way, for the seaward margin of Downholland Moss was overblown by sand a little after 4090±170 BP (Table 1, No. 63).

Sweden

There is closer agreement between the temporal pattern of marine transgressions established by Mörner (1969a) along the west coast of Sweden and the pattern established in north-west England. Mörner has defined ten marine transgressions during the Flandrian stage based on morphological and stratigraphical evidence. He has established the age of the transgressions by radiometric and pollen analyses. The evidence derives in part from detailed stratigraphical descriptions from the Viskan valley, in part from a reinterpretation of von Post's work in the Viskan valley (von Post 1968), and in part from a detailed examination of buried, elevated shorelines along the Swedish west coast. The following time limits are given by Mörner for the marine transgressions along the west coast of Sweden:

	Mörner (1969a)		Mörner (1976a)	
ALV-2	9280 to 8850	years BP	7750 to 6900	years BP
PTM-2	7050 to 6900	years BP	6450 to 6200	years BP
PTM-3	6450 to 6250	years BP	5900 to 5500	years BP
PTM-4	5900 to 5550	years BP	5000 to 4550	years BP
PTM-5	4975 to 4550	years BP	4300 to 3950	years BP
PTM-6	4250 to 3950	years BP	2700 to 2400	years BP
PTM-7	3600 to 3200	years BP	2300 to 2200	years BP
PTM-8	3050 to 2900	years BP	1800 to 1300	years BP
PTM-9	2650 to 2200	years BP	1100 to 1000	years BP
PTM-10	1200 to ?	years BP		

Ancient Lake Veselången—2(ALV-2) represents the First Flandrian transgression in Sweden, and began about 9280 years BP and reached a maximum about 8850 years BP. Post-glacial Transgression Maximum Level—2(PTM-2) is well marked in the Viskan Valley at 7750 years BP, but the maximum levels were attained later from 7050 to 6900 years BP.

There is a very close correlation between ALV-2 and Lytham I. The transgressive phase of Lytham I is registered in Morecambe Bay at about 9270 years BP and the culminating phase at Heyhouses Lane, Lytham, at 8575 years BP. The onset of Lytham II at 8390 years BP is well marked at several sites in the Fylde, but there is no evidence for a transgression with such time limits in Sweden; indeed, Lytham II corresponds with ALV-3 from about 8500 until 7750, which in Sweden is a period of marine regression, when Ancient Lake Veselången was a freshwater lake, according to Mörner.

The age for the maximum limits of PTM-2 are out of phase with Lytham III from 7605 to 7200 years BP, but when the onset of PTM-2 in the Viskan Valley at 7750 years BP is considered, then PTM-2 and Lytham III constitute an expression of a late Flandrian I transgression in North-west Europe. PTM-3 falls well within the time limits established for Lytham IV. However, at Eskilstorp, where Mörner has recorded the transgressive and regressive contacts of PTM-3, the onset of PTM-3 is dated at 6520±105 years BP, which is closer to the inception of marine sedimentation constituting Lytham IV. Furthermore, at Höganäs, where PTM-3 is also registered, the opening stages of the transgression are dated at 6540±100 years BP. There is a close correlation between PTM-4 and Lytham V, but there is little evidence of synchroneity between PTM-5 and Lytham VI. The time limits for PTM-5 have been established from a sampling site at Torekov (B676): here an assay on material immediately subjacent to marine sand gave a date of 4890±100 years BP, yet the AT-2/SB-1 pollen zone boundary is drawn at a level well within the sandy stratum, and there is a date for this pollen zone boundary of 5050 years BP. There are no pollen analyses on the peat from which the radiocarbon date was obtained, but there can be little doubt that this date is too young. In addition, the culminating stage of PTM-5 is dated at 5025±100 years BP, which is very close to many of the dates obtained on the regressive contact of Lytham VI. There is close agreement on the time limits for PTM-7 and Lytham VII (Mörner 1969a), but there is no evidence in north-west England for PTM-6; elsewhere in England, however, there is evidence for this transgression in the Fenlands and the Somerset Levels. The onsets of PTM-8 and of Lytham VIII occur more or less simultaneously, but the culminating stages occur later in north-west England than in Sweden. Evidence for the culminating stages of PTM-8 comes from the Viskan valley, where there are no radiocarbon dates for this stage. PTM-9 falls within the time limits established for Lytham VIII. Mörner's (1976a) revised time limits for PTM-2 to -10 confirm the concurrence of PTM-2 to -5 and L I to VI. PTM-7 to -10 correlate with L VIII to X, but L VIII now includes two transgressions in Sweden, PTM-7 and -8.

In Blekinge, south-east Sweden, Berglund (1971) has identified six marine transgressions based on detailed stratigraphic, pollen and diatom analyses, particularly from Siretorp and Hallarums Mosse. The time limits for the transgressions are given below; together they constitute stages in the restoration of sea-level in the Baltic during the Littorina Stage.

Transgression I	Early Atlantic	6950 to 6650 years BP
Transgression II	Mid-Atlantic	6450 to 6250 years BP
Transgression III)	Late Atlantic	5850 to 5550 years BP
Transgression IV)		5450 to 5250 years BP
Transgression V	Early Sub-Boreal	4650 to 4450 years BP
Transgression VI	Middle Sub-boreal	4050 to 3850 years BP

Digerfeldt (1975) has identified seven marine transgressions from the former lagoon of Barsebäcksmossen in western Skåne, Sweden, and correlated them with the transgression sequences of Berglund and Mörner. In comparison with the marine transgression sequences established in north-west England, some of the transgressions recorded at Barsebäcksmossen occur within their time limits, whereas others are out of phase. For example, Lytham V 4984-4785 B.C. (T. corr. T$\frac{1}{2}$5730) occurs during the latter part of the first marine transgression at Barsebäcksmossen dated 5200-4800 B.C. Transgressions 2 and 3, occur during the complex sea-level episode Lytham VI from 4579-3803 B.C.: sea-level maxima during this episode are dated at 4313 B.C. and 3981 B.C. and coincide with a regression between Transgressions 2 and 3, and the end of Transgression 3. Lytham VIa 3331-2958 B.C. comprehends Transgression 6 and Lytham VII 2284-1546 B.C. comprehends

Transgression 7. The existence of a bar at the seaward edge of the former lagoon may have delayed the registration of marine events at Barsebäcksmossen, making correlation difficult.

Some of Berglund's, Mörner's and Digerfeldt's dates have been corrected by up to 300 years to allow for the delayed transmission of carbon to marine sediments; this sea correction may be as high as 305±25 years (Mörner 1969) on the Swedish West Coast or as low as 60±90 in the Baltic Sea (Berglund 1971). No correction has been attempted on dates from north-west England, even though some of the dated samples were brackish water peats and muds. For this reason, the data are not strictly comparable, and time displacements of several hundred years can be expected. Nevertheless, the temporal synchroneity of the marine transgression sequences across north-west Europe is close in many cases (Fig. 40) and undoubtedly reflects eustatic movements of sea-level. The actual time of the onset and culmination of a marine transgression reflects eustatic movements, but conditioned by local site factors, so that a consideration of tectonic and isostatic factors becomes important. In addition, a single transgression will have a range of dates establishing its transgressive and regressive phases, and the distribution of these dates is not necessarily normal, but may be skewed. The establishment of time limits on a transgression usually derives from a single site, which may lie eccentrically within the age limits of the transgression. Notwithstanding these difficulties, the foregoing description of marine transgression sequences in North-west Europe permits a tentative correlation with sequences in north-west England. It is hoped that refinement of the time limits will permit greater confidence in establishing homologues and stronger European correlations.

2. RELATIVE SEA-LEVEL CHANGES

2.1 EVIDENCE FROM BRITAIN

Ten relative sea-level curves have been published using data from the British Isles, and are shown on Fig. 41. Three of the sea-level curves derive from areas recording isostatic uplift: in the Forth valley (Graph A on Fig. 41, Sissons 1966, 1967; Sissons and Brooks 1971), in Southern Scotland (Graph B on Fig. 41, Donner 1970), and in Cumbria (Graph D on Fig. 41, Andrews *et al.* 1973). Five of the curves derive from areas undergoing subsidence: in the Fenlands (Graph I on Fig. 41, Godwin 1940a), in the outer Thames estuary (Graph H on Fig. 41, Greensmith and Tucker 1973), in the English Channel (Graph G on Fig. 41, Clarke 1970) and in the Bristol Channel (Graph E on Fig. 41, Hawkins 1971; Graph F on Fig. 41, Kidson and Heyworth 1973). The balance of the sea-level curves derive from intermediate areas that record uplift immediately after the end of the Devensian glaciation, followed by slight downwarping, which either continues at the present time, according to Valentin's (1953) analysis of tide gauge data or may have ceased; these two sea-level curves derive from data from west Lancashire, described earlier

(Graph C on Fig. 41), and from north-east England (Gaunt and Tooley, 1974).

RELATIVE SEA-LEVEL CURVES FROM THE BRITISH ISLES

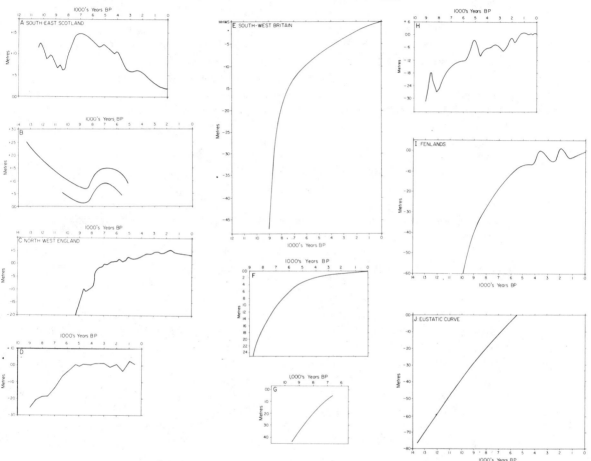

FIG. 41. Relative sea-level curves from the British Isles. Three of the curves, A, B and D (Sissons 1967, Donner 1970 and Andrews *et al.* 1973) derive from areas undergoing isostatic recovery, five curves from areas undergoing downwarping, E, F, G, H and I. (Hawkins 1971; Kidson and Heyworth 1973; Clarke 1970; Greensmith and Tucker 1973; Godwin 1940) and one curve C (north-west England) from an area that has recorded no vertical movements since 5500 years B.P. The eustatic sea-level curve of Godwin, Suggate and Willis (1958) J is added, for it includes index points from British coasts.

Sissons (1967) and Sissons and Brooks (1971) have constructed sea-level curves from an area east of the Menteith moraine in the Forth valley based on the age and altitude of raised shorelines. A period of relatively high sea-level is recorded in both the Forth valley and west Lancashire during the middle of Flancrian Ic: Lytham I and the formation of the Low Buried Beach appear to have occurred more or less at the same time. A relative fall in sea-level occurred in west Lancashire at 8575±105 BP (Hv. 4346) and in the Forth valley at 8690±140 BP (I.1839). A subsequent rise of sea-level and the deposition of Carse clay in the Forth valley is dated at 8270±140 BP (I.1838) and falls within the period of Lytham II. The rapid rise of sea-level recorded in the Forth valley between 8270 and 7480 radiocarbon years ago, and constituting Sissons' Major Post-glacial Transgression (PG1), occurred during a period of rapid rise of sea-level in west Lancashire. Further oscillations of sea-level have been recognised in the Forth valley: the PG 2 shoreline appears to be synchronous with Lytham VI; there is no homologue of PG 3 in west Lancashire and PG 4 probably occurred during Lytham VIII.

The curves of relative sea-level changes drawn up by Donner (1970) from data from central and south-west Scotland show a falling sea-level until about 9000 radiocarbon years ago, after which sea-level rose rapidly to a maximum about 7000 years ago and subsequently fell. It is difficult to compare this curve with the sea-level curve from west Lancashire because the index points used by Donner are few and some are of suspect provenance; furthermore, they derive from two areas on the south-east and south-west coasts of Scotland.

Further south, in Cumbria, Andrews *et al.* (1973) derive a eustatic sea-level curve from a consideration of ten index points. Their relative sea-level curve shows an initial period of rising sea-level between 9200 and 6000 radiocarbon years ago, followed by three periods of relatively high sea-level about 3200, 2000 and 900 radiocarbon years ago. In west Lancashire, between 9200 and 6000 radiocarbon years ago, four periods of relatively high sea-level have been recorded, Lytham I-IV: the high sea-levels recorded subsequently appear to be associated with Lytham VII and X. There is no evidence for a period of relatively high sea-level at 2000 radiocarbon years ago in west Lancashire. There is some concern over the provenance of the dated samples used as index points for the sea-level curve, not only because they comprise a mixture of materials—shell samples, wood, peat and charcoal —but also because the samples are not related unequivocally to former positions of sea-level (see Huddart and Tooley 1972, Tooley 1977d).

In north-east and eastern England, Gaunt and Tooley (1974) have described sixteen index points and their relationship to former positions of sea-level. The oscillations of M.H.W.M.S.T. that these index points show are in phase with the sea-level oscillations from west Lancashire, but are displaced in altitude.

Further south, in the Fenlands, Godwin (1940a) published the first curve showing relative sea-level changes during the Flandrian stage in England, based on 34 index points. The chronology was based on pollen analytical and archaeological evidence. Godwin's curve shows a rapidly rising sea-level relative to the land from the opening of the Flandrian stage until about 5000 years BP. During the first millennium of Flandrian III there appears to have been a still-stand. The first part of the curve from the Fenlands appears to correspond with Lytham I-V, although the evidence considered on pp. 157–162 indicates that the Fenlands were not subject to extensive marine transgression until Lytham VI. The discrepancy arises in part because Godwin's curve has not been calibrated radiometrically and in part because the deep peats in the Fenlands, from Guyhirn, for example, have not been subject either to detailed pollen analysis or to radiocarbon dating. The still-stand covers that period for which there is evidence in north-west England and elsewhere of a fall in sea-level. Between 4000 years BP and the present, Godwin shows an oscillating sea-level, rising above present sea-level about 3700 and 2000 years BP and probably corresponding to Lytham VII and Lytham VIII, both of which rose above present sea-level. Correlation with the oscillations established by Godwin in the Fenlands would be stronger if radiocarbon dates were available from each of the sites contributing index points on Godwin's graph.

Although Godwin's curve shows an oscillating relative sea-level from 5000 years BP to the present, the eustatic curve (Graph J on Fig. 41 Godwin *et al.* 1958) shows a continuously rising sea-level, actively under way during Flandrian I, and terminating during Flandrian II at about 5500 years BP. Jelgersma (1961) argued that at the end of Flandrian II, sea-level in the Netherlands stood 5 metres below its present level. This area is affected by subsidence, and the figure is too great. In north-west England, Mean Sea Level was a little over a metre below its present level, and, in Sweden, Mörner shows sea-level about two metres below its present level, some 5000 years BP. Although sea-level must have been close to its present level during the FII/FIII transition, there is ample evidence that it oscillated subsequently. This conclusion is supported by Greensmith and Tucker (1973), who consider evidence from the outer Thames estuary. They describe six marine transgressions during the Flandrian stage, associated with periods of relatively high sea-level. Transgression I is correlated with Lytham I and Transgression II with Lytham III-VI; the sustained transgression in the outer Thames estuary while four distinctive transgressions were recorded in west Lancashire may be an expression of continuing subsidence in south-east England within the southern North Sea Basin. Transgression III and Lytham VII appear to be homologous. Transgressions IV, V and VI in the outer Thames estuary may be associated with the periods of relatively high sea-level recorded in west Lancashire as Lytham VIII, IX and X. The lack of detailed concurrence derives from the fact that Greensmith and Tucker have used dates on peat, shells and shelly silt to derive the index points on their relative sea-level curve: the peat samples do not appear to have been related closely to their contemporary sea-level and the shell samples in all but one case derive from allochthonous deposits, and the dates have not been corroborated independently.

Further up the Thames, Devoy (1977) has identified five marine transgressions and the periods of relatively high sea-level with which they are associated. During the relatively high sea-level of Thames I, two periods of high sea-level were recorded in north-west England (Lytham II and III). During Thames II, Lytham IV and V were recorded, and Lytham VII during Thames III. Thames IV is associated with Lytham VIII and Thames V with Lytham IX.

In south-west England, Hawkins (1971) has plotted 41 variates on a time/depth graph. Dated samples are derived in part from radiometric assays and in part from pollen analyses referred to the numerical pollen zones established by Godwin for Britain. All samples are plotted in relation to the High Water of Spring Tides, but there is no indication whether a single altitude or several altitudes have been used, as samples from the English Channel, Bristol Channel and Severn estuary are all plotted on the same graph, so that comparison with other sea-level curves is impossible. Furthermore the rationale for the inclusion of many of the variates is obscure and requires elucidation. The four radiocarbon dates from Port Talbot are plotted, and three of them certainly 'do not conform with the suggested sea-level pattern', because the assays were intended to establish pollen assemblage zone boundaries: Godwin and Willis (1964) are quite explicit about this point (p. 124). Of the six dates from the Port Talbot cores, only one (Q.663) dates the onset of marine conditions in this area, and this date may be maximal: 'Sample Q.663 indicates that the postglacial eustatic rise of sea-level above −62 ft. O.D. on this coast must have taken place after 8970±160 BP, or *at* this date if no depositional gap exists between the top of this peat bed and over-lying marine deposits.' The use of the Gordano material on the sea-level curve is inadmissible; the evidence that Jefferies *et al.* (1968) present is also quite explicit: biogenic sedimentation appears to have proceeded uninterrupted in part of the valley during the Late Weichselian and Flandrian stages, and only the eastern half of the valley was affected by one or possibly two of the transgressions constituting the Flandrian restoration of sea-level. It has been suggested elsewhere (p. 165) that the Gordano valley was probably affected by a late Flandrian II/early Flandrian III transgression which is a homologue of Lytham VI. The regressive contact of this transgression is not shown on Hawkins' graph. Several dates derive from inter-tidal peat beds in the Bristol Channel, where the relationship to sea-level at the time of biogenic sedimentation does not appear to have been established. The presence of an intertidal peat bed, a so-called 'submerged forest', does not permit the establishment of a former position of sea-level axiomatically. Many peat beds recorded in the intertidal zone and in cliffs at the head of a contemporary marine beach are found there as a result of contemporary coastal processes, isolating the basal sediments of kettle holes, (for example, Glanllynnau, Caernarvonshire (Simpkins 1974), Rossall Beach (Tooley 1969, 1977a) and Allonby (Huddart and Tooley 1972)) and eroding the circumvallating till. Godwin made this point also, which seems to have gone unheeded: 'It is even possible, in some circumstances, for a submerged peat bed to appear upon our coastline between tide marks, without any relative shift of sea- and land-level: this may occur where shore erosion exposes the fresh water muds and peats of lake or river systems at low level behind the coast' (Godwin 1956, p. 24). A-propos the dated intertidal peats that have been used on Hawkins' graph, those from Freshwater West (Q.530) and Westward Ho! (Q.672) do not appear to be related to their contemporary sea-levels. In the case of the intertidal peat at Stolford, shown on Hawkins' graph, Kidson (in Callow and Hassall 1968) has established the relationship of the dated levels to former positions of sea-level during Flandrian II and III, and such levels serve as incontrovertible index points on any sea-level curve.

Kidson and Heyworth (1973) have described the rise of sea-level during the Flandrian stage, using twenty-five variates from sites on both sides of the Bristol Channel. The recorded altitude of each variate has been adjusted to take into account gravitational compaction since the formation of the deposit, and the age of each variate corrected from radiocarbon years to sidereal years, according to the Bristlecone pine calibrations of Suess (1967, 1970a) and Ferguson (1970). The difficulties of correcting the altitude of dated samples, recognised by Kidson and Heyworth, are discussed in Chapter 1, but the oscillation in sea-level shown on their uncorrected curve (Fig. 7, 1973) is inexplicable in terms of either consolidation or age correction. Furthermore, as the radiocarbon time scale can be adjusted from radiocarbon years to sidereal years with some confidence between 500-2200 radiocarbon years ago (Damon et al. 1973) and with less confidence between 4800-6500 radiocarbon years ago, it is difficult to appreciate what the age correction on Kidson and Heyworth's sea-level curve has achieved, particularly as nine of their radio-carbon dates fall within the periods of low confidence limits and three of their dates (I.4402, I.4403 and Q.663) lie outside the range of dendrochronologic calibration, but have, none the less, been 'corrected'.

A basic assumption made by Kidson and Heyworth (1973) in presenting their eustatic sea-level curve from the Bristol Channel is that this area is stable. It has not suffered either positive or negative movements for at least the last 10,000 years, and has remained stable for much of the Quaternary (Kidson 1977). They dismiss Walcott's theoretical considerations (1972) that the effects of ice loading will be recorded up to 1500 km beyond the ice limits of the last glaciation. However, the empirical studies of Valentin (1953), Rossiter (1972) and Lennon (1975) indicate a downward movement throughout southern England and along the French coast. Rossiter has shown that the relative rise of sea-level at Newlyn during this century is 2.2mm/year, and, given a eustatic trend of +1mm/year, Newlyn is subsiding at the rate of 1.2mm/year. Lennon (1975) identified a nodal line running along the

Latvian coast across Denmark and the North Sea to the Humber; uplift north of this line is occurring with maximum values of +8mm/year in the Gulf of Bothnia, and, 'to the south and west, motion in the opposite sense is indicated with land elevation falling at a rate of approximately 1mm/year in the whole of southern England' (p. 9). It is difficult to conceive that these epeirogenic movements are twentieth-century phenomena. They are long term and affected southern England for at least the period of this interglacial. The considerable thickness of Flandrian sediments in the Bristol Channel and Somerset Levels bears witness to subsidence in these areas, as also in the Lower Thames Valley and the Netherlands, where West (1972) and Jelgersma (1961) have demonstrated downwarping during the Quaternary. Although Gaunt and Tooley (1974) have demonstrated downwarping during the Flandrian stage in eastern England, West (1972) has concluded that in south-east England 'there is no clear geological evidence that subsidence has played a significant part in determining the course of sea-level rise' during the Flandrian stage. Devoy (1977) however has indicated that the history of subsidence in south-east *and* south-west England is similar, and that the south-west has been subject to downwarping. Clearly there is a need for future research to resolve these conflicting interpretations of the evidence, particularly for south England during the Flandrian stage.

Of the relative sea-level curves considered from the British Isles, there is a fundamental distinction between authors arguing for a continuously rising sea-level during the Flandrian stage, and those arguing for an oscillating sea-level. The evidence presented in Chapter 4 for an oscillating sea-level in north-west England is supported by Godwin from the Fenlands, Gaunt and Tooley from north-east England, Greensmith and Tucker in the outer Thames estuary and by Sissons from the Forth valley. The absence of oscillations on the relative sea-level curves of Hawkins and Kidson and Heyworth derives from their interpretation of the index points used on their sea-level graphs.

2.2 EVIDENCE FROM NORTH-WEST EUROPE

During the last fifteen years many sea-level curves have been drawn, based on detailed stratigraphic analyses in coastal areas of north-west Europe. From different parts of the European coasts, recording either downwarping or uplift, Mörner (1973) has argued that the Brittany coast of France is probably the most stable, and eustatic movements of sea-level will be recorded there directly at successively higher altitudes and need not be inferred from unstable areas after a necessary transformation of the numerical data. However, if Walcott's (1972) thesis is accepted, then this area and coastal areas further south will be adjusting to the unequal crustal loading of the Quaternary glaciations in this part of the northern hemisphere, by subsiding in response to uplift further north. The long-period, relaxation time involved, together with regional factors such as subsidence within the North Sea basin and uplift in

Scotland and Fenno-Scandinavia, makes it impossible to identify a stable coastal unit where eustatic changes are registered directly. For this reason, all sea-level curves generated from data gathered around the coasts of North-west Europe, including the British Isles, are regarded as a reflection of relative sea-level changes, and are plotted as such on Figs. 41 and 42.

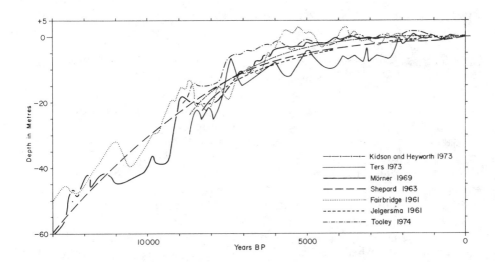

FIG. 42. Relative sea-level curves for the last 13 000 years from the Swedish west coast (Mörner 1969a), the Netherlands (Jelgersma 1961, 1966), the French coast (Ters 1973), north-west England (Tooley 1974), the Bristol Channel (Kidson and Heyworth 1973), North-west Gulf of Mexico (Shepard 1963) and the world (Fairbridge 1961). The curve from north-west England has been dropped by 4.0 metres, so that it approximates the relative change in Mean Tide Level (cf. Fig. 38).

In 1961, Jelgersma published a detailed description of evidence for sea-level changes in the Netherlands. Three curves showing relative sea-level movements in different parts of the Netherlands were drawn, and a fourth curve (reproduced in Fig. 42) showing the movement of Mean Sea Level during the Flandrian stage in the Netherlands was estimated in 1966. The curve of Mean Sea Level has been drawn a metre below Curve I (1961), which records the age and depth at which biogenic sedimentation began on the sloping Pleistocene subsurface; the fresh ground water table is then equated with the high tide level. The curve has been smoothed, and shows a continuously rising sea-level, decreasing after 6000 years BP. Curve II (1961), based on samples from the Rhine–Meuse estuary, shows a rapid rise in sea-level until 6000 BP, a still-stand between 6000 and 5600 years BP, a further rise in sea-level until 5000 years BP, followed by a still-stand and a slow rise after 4000 years BP. Jelgersma argues that fluctuations of sea-level can be explained in terms of actual and inherent errors in the variates plotted on the time-depth graph such as consolidation, tidal inequalities and changing river discharges. This argument is accepted generally in the Netherlands, and the evolution of the Dutch coast is explained in terms of a continuously rising sea-level during the Flandrian stage, the rate of rise slackening as the present is approached (Hageman 1969).

None of the variates plotted on Jelgersma's time-depth graph is directly related to a former sea-level, but the approaching marine conditions are inferred from rising freshwater tables and the inauguration of biogenic sedimentation. In many cases, as explained in Chapter 1, the actual date of arrival of marine sedimentation may be more than 700 radiocarbon years after biogenic sedimentation began and at least a metre higher. The altitudes of the index points upon which the curve is based, could be raised and the timing advanced by these values. If forty-one radiocarbon dates from the transgressive and regressive contacts of transgressions identified in the Netherlands are plotted, a curve drawn through the lowest points shows that an oscillating sea-level can be derived from the Dutch data. An oscillating mean high water curve has been constructed by Kooijmans (1974) for the Rhine–Meuse delta using archaeological and stratigraphic evidence. Although the general trend of Kooijman's sea-level curve accords with Jelgersma's curve, it differs in detail in showing persistent, well-marked oscillations of high water mark.

A sea-level curve from the Atlantic coast of France (Delibrias and Guillier 1971) is also smoothed, and shows a continuously rising sea-level, the rate of rise decreasing from 7000 years BP to the present. During the period from 2000 to 500 years BP, sea-level rose slightly above present sea-level, and this period comprehends Lytham IX during which sea-level in north-west England appears to have risen 1.8 m above present sea-level. The sea-level curve of Delibrias and Guillier is quite distinctive, compared to the oscillating sea-level curves of Morzadec-Kerfourn (1974) for the north-west coast of France and Ters (1973) for the Atlantic coast of France (Fig. 42). Madame Ters has established seven periods of relatively high sea-level which are closely correlated with periods of relatively high sea-level in west Lancashire. The oscillations of sea-level that occurred between 8200 and 7900 radiocarbon years ago, and defined as *Les niveaux du Havre*, occurred during Lytham II. *Le haut niveau de Fromentine* occurred during Lytham III and *Le haut niveau de Bréhec* during Lytham IV. The evidence from Tréopan indicates a low sea-level at a time during the final millennium of Flandrian II when a transgression, Lytham V, was under way in north-west England. Lytham VI is closely associated with *Le haut niveau de Bretignolles* but Lytham VIa occurs during a period of relatively low sea-level, recorded at Argenton. The succeeding three marine transgressions and periods of high sea-level recorded in west Lancashire, Lytham VII, VIII and IX occur during periods of high sea-level in France, defined as *Le haut niveau de Camiers, Le haut niveau de Champagne* and *Le haut niveau de Saint Firmin*. Difficulties in comparing the sea-level curves from the Atlantic coast of France with those from west

Lancashire derive from differences in scale: the sea-level curve from west Lancashire came from a restricted and homogenous area, whereas the curve from France derives from a consideration of 160 index points extending from Calais to Aunis—some 600 km. Furthermore, the amplitude of the oscillations is considerable and would require massive additions to and abstractions from the water in the ocean basins, for which there is no evidence from high latitude glaciers and ice caps during the Flandrian stage of commensurate increases and decreases in ice volumes.

Further north, in Sweden, there are similarities in the form of the curves of relative sea-level produced by Berglund (1971) and Mörner (1969) and the curve from west Lancashire. The relative sea-level curve from southern Sweden shows an oscillating sea-level between 7000 and 3850 years BP. Transgression I in Southern Sweden occurred during a period when relative sea-level rose to about +6.5 metres above present sea-level, and brackish to marine sediments were laid down in Hallarums Mosse, Siretorp, Förskjob. This early Flandrian II transgression corresponds to Lytham IV, when mean sea-level rose to −4.1 metres O.D. A second transgression in Southern Sweden during which relative sea-level rose to +6.8 metres above the present level corresponds to an oscillation in sea-level, during Lytham IV, when sea-level rose to −3.1 metres O.D. and then fell slightly. Transgression 3 occurred between 5850 and 5550 years BP, during which relative sea-level rose steeply in Southern Sweden to about +7.6 metres and this corresponds to an oscillation in sea-level in north-west England, when sea-level rose to -2.0 metres O.D., during Lytham VI. Transgression 4 is not well represented in north-west England, although there is evidence of a slight oscillation during a period of rising sea-level. There is no evidence for Transgression 5 in north-west England, when Berglund argues for a high relative sea-level at about +5.8 metres, and the onset of Transgression 6 appears to be advanced in Southern Sweden.

Mörner (1969) has derived a sea-level curve from the west coast of Sweden by transforming height data from raised shorelines, and arguing that absolute changes in sea-level can thus be calculated. The sea-level curve from the Swedish west coast is displayed in Fig. 42. The patterns of Mörner's curve and of the sea-level curve presented here are remarkably similar, the more so because the procedures of derivation of the curves are different and the curves have been established in areas with different tectonic and glaciation histories. The sea-level curve from north-west England has been derived without any numerical transformation of the data, except for the simple graphical device of reducing a curve showing the movement of Mean High Water Mark of Spring Tides to a Mean Sea Level curve by deducting 4 metres—the contemporary difference between M.H.W.S.T. and M.S.L. on the Lancashire coast at St. Annes-on-the-Sea. Between 9000 and 5400 years BP the sea-level curves are asymptotic: between 5400 and 2000 years BP they anastomose, and their mean altitude probably reflects eustatic changes. If Mörner's curve registers eustatic changes, then the divergence of the two curves during Flandrian I and II can be explained by continued isostatic recovery in the type area of north-west England, for which no allowance has been made on the curve. It would follow then, that in the type area since 5400 years BP no uplift has taken place, and that south-west Lancashire has registered largely eustatic changes directly. If the curve of Mean Sea Level from north-west England between 9000 and 5400 years BP is adjusted by the difference in height between the transgressions registered in south-west England, Wales and the type area (i.e. if the differences in height given in Chapter 3 are deducted from the index points on the sea-level curve), then Mörner's curve and the curve from north-west England coincide also for this period. The conclusion from these two sea-level curves is that their over-all trends reflect eustatic changes in sea-level and that slight differences after isostatic effects are taken into account reflect local site conditions for which no allowance can be made.

3. CLIMATIC IMPLICATIONS

In 1961, Fairbridge demonstrated a close correlation between minor os-cillations of sea-level and climatic fluctuations. Jelgersma (1966), whilst accepting climatic fluctuations, argued strongly against sea-level oscillations on the basis of the inadequacy and insufficiency of the data. The causal relationship explicit in Fairbridge's correlation was thereby refuted by Jelgersma. Some authors (Mörner 1969a, 1969b) have supported Fairbridge's general conclusions, whereas others (Kidson and Heyworth 1973; Scholl and Stuiver 1967; Shepard 1963; Shepard and Curray 1967; Jelgersma 1966) have advanced evidence against them, or imply criticism.

The evidence presented in Chapter 4 indicates that pronounced sea-level oscillations have occurred: low amplitude oscillations undoubtedly register local effects, such as consolidation, and high amplitude oscillations include a large eustatic component.

Since 1961, evidence from a range of sedimentary environments, both terrestrial, limnic and marine, has proved beyond doubt the existence of climatic fluctuations during the Flandrian stage and has given greater pre-cision to their magnitude, direction and age.

It remains to establish unequivocally the relationship between an oscillating sea-level and climatic shifts during the Flandrian stage, thereby reinforcing the conclusions of Fairbridge (1961), Fairbridge and Hillaire-Marcel (1977) and Mörner (1969a), or to refute the relationship by demonstrating a lack of concurrence between the two global phenomena.

The oscillations on the sea-level curve, derived from data from north-west England, are associated with eleven marine transgressions, of which three — Lytham III, IV and VI — include oscillations of low amplitude. These derive from local effects, which none the less do not invalidate the general trend recorded by the transgressions. Periods of marine transgression are associated with positive eustatic changes, and regression with negative eustatic

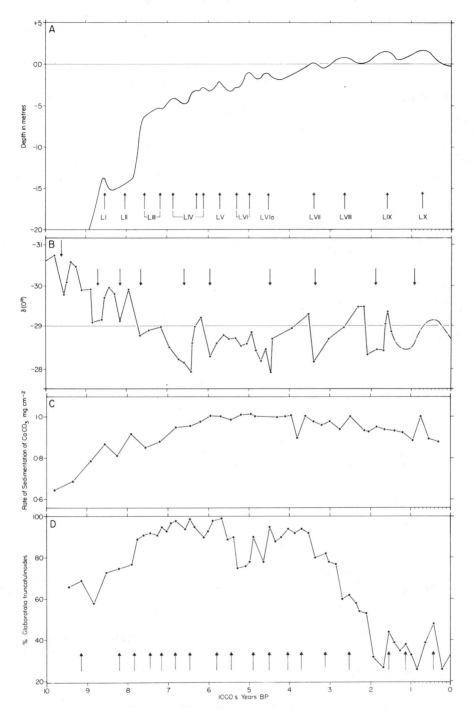

FIG. 43. (A) Sea-level oscillations in north-west England. (B) Climatic variations inferred from the relative deviation of the O^{18}/O^{16} ratio from that of Standard Mean Ocean Water at Cape Century, Greenland, both according to Dansgaard *et al.* (1969). (C) Changes in the rate of Ca CO_3 sedimentation from a mid-Atlantic Equatorial core, according to Wiseman (1966). (D) Change in the frequency of left-coiling tests of *Globorotalia truncatulinoides* in a core from the Atlantic, according to Wollin *et al.* (1971), from which climatic variations can also be inferred. Arrows on graph (A) indicate periods of relatively high sea-level attained during transgressions Lytham I-X: on (B), periods of ice attenuation, glacier retreat and inferred climatic amelioration: on (D), peak frequencies of *G. truncatulinoides* and inferred periods of climatic amelioration.

changes. The sea-level curve from north-west England has been re-plotted on Fig. 43(A) and the arrows indicate transgression maxima. These maxima are closely correlated with oscillations in the $\delta(^{18}O)$ isotope curve derived from the Cape Century ice core also shown on Fig. 43(B), from which climatic events can be inferred (Dansgaard et al. 1969). Periods of climatic amelioration in Greenland appear to comprehend periods of relatively higher sea-level and marine transgression episodes in north-west England. A similar close correlation between sea-level oscillations and the $\delta(^{18}O)$ isotope curve, has been established by Ters (1973).

Further, climatic indicators from different sedimentary environments tend to reinforce these correlations.

In 1966, Wiseman published a graph showing the changing rates of sedimentation of $CaCO_3$ from an Equatorial Atlantic Ocean core (01°10'N.19°50'W). Relatively high rates of carbonate sedimentation were related to increased temperatures of the ocean surface and reduced rates to decreased temperatures. Short-term oscillations were superimposed on long-period trends, both of which were correlated with climatic oscillations inferred from terrestrial environments of higher latitudes. The following chronology for high rates of carbonate sedimentation corresponds to periods of marine transgressions in north-west England:

P. 35 at 8550 yrs BP is associated with Lytham I	9270–8575 yrs BP
P. 33 at 7900 yrs BP is associated with Lytham II	8390–7800 yrs BP
P. 30 at 6800 yrs BP is associated with Lytham IV	6710–6157 yrs BP
P. 27 at 5950 yrs BP is associated with Lytham V	5947–5775 yrs BP
P. 23 at 4950 yrs BP is associated with Lytham VI	5570–4897 yrs BP
P. 16 at 3600 yrs BP is associated with Lytham VII	3700–3150 yrs BP
P. 11 at 2500 yrs BP is associated with Lytham VIII	3090–2270 yrs BP
P. 6 at 1400 yrs BP is associated with Lytham IX	1795–1370 yrs BP

Furthermore, the relatively low sea-level between Lytham VI/VII is associated with P.17 at 3800 radiocarbon years, when the rate of carbonate sedimentation was much reduced.

Wiseman's graph of carbonate sedimentation and the graph of $\delta(^{18}O)$ isotope are based on assumed accumulation rates derived from empirical models, upon which their chronologies depend. Neither is strictly comparable with the sea-level curve from west Lancashire, the index points of which are based on uncorrected radiocarbon dates. Mörner (1972) has drawn attention to the chronological anomalies of the $\delta(^{18}O)$ curve of Dansgaard et al. (1969). This objection, however, is partly eliminated when the sea-level curve is compared with the curves described by Wollin et al. (1971) and taken from ocean cores from the Atlantic and Pacific oceans, which are supported by radiocarbon dates.

Core A179-15 (Wollin et al. 1971) from the Atlantic ocean (24°48'N. 75°55'W) shows the change in frequency of *Globorotalia truncatu-*

linoides, and temperature oscillations of surface ocean waters at this latitude can be inferred from the frequency changes of taxa constituting the
foraminiferal assemblages at successive levels in the core. During the
Flandrian stage, the general increase in temperature until 5500 years BP is
recorded, followed by a decline from about 3300 years BP to the present. An
oscillating pattern is superimposed on this general trend. Both the over-all
pattern and the detailed oscillations of Wollin *et. al.*'s curve tend to support
the sea-level curve from north-west England.

Increases in the frequency of *G. truncatulinoides* in the curve of Wollin *et
al.* from the Atlantic Ocean core are associated with marine transgressions in
north-west England. Lytham I and II occur during a period when the
frequency of *G. truncatulinoides* rises from about 58 to 90 per cent between
100 and 108 cm. Lytham II occurs at a time when the foraminiferal frequency
increases slightly between about 8624 and 7895 years BP. (Dates have been
calculated by interpolation from Fig. 4 in Wollin *et al.* 1971.) Lytham III
correlates with a peak frequency of *G. truncatulinoides* about 7200 years BP
at 90 cm; Lytham IV with a peak about 6300 years BP at 80 cm; Lytham VI
with a peak about 5300 years BP; Lytham VII with a peak at 3780 years BP;
Lytham VIII with a peak about 2520 years BP; and Lytham IX with a peak
about 1470 years BP.

Denton and Karlén (1973) and Karlén and Denton (1976) have drawn
attention to the close correlations of glacier variations during the Flandrian
stage and fluctuations in atmospheric ^{14}C; glacier advance is associated with
increased atmospheric ^{14}C and retreat with decreased atmospheric ^{14}C. Short-
term variations in atmospheric ^{14}C are closely associated with advance and
retreat stages of the Little Ice Age in the Sarek and Kebnekaise Mountains in
northern Sweden. However, earlier glacial events during the Flandrian do not
show the same detailed variations as the variations in atmospheric ^{14}C would
imply. Using the data of Suess (1970a, b) and Denton and Karlén (1973), it
can be shown that a close relationship exists between periods of increased
atmospheric ^{14}C and low sea-levels in north-west England. Periods of
increased atmospheric ^{14}C occur at about A.D. 700, 400 B.C., 1400 B.C., 3160
B.C., 3650 B.C. and 4200 B.C., *inter alia*, and these compare with periods of low
sea-level in north-west England about A.D. 630, 432 B.C., 1501 B.C., 2964 B.C.,
3556 B.C. and 4292 B.C..

Using 330 variates, Geyh and Streif (1970) demonstrated that sea-level
changes were synchronous along the southern North Sea coast. Furthermore,
Geyh (1971) showed that by correcting ^{14}C dates to calendar years for coastal
peat formation along the southern North Sea coast there was a close
relationship with variations in atmospheric ^{14}C: periods showing maxima on
the ^{14}C curve are related to periods of coastal peat formation, and periods
showing minima on the ^{14}C curve to periods of marine clay and silt sedimentation associated with rises of sea-level. Of the fourteen transgression
episodes identified by Geyh (1970) from the southern North Sea coast, seven

can be recognised in north-west England, thereby supporting Geyh's conclusion that 'from the chronostratigraphy of the marine Holocene the course of a global geological process can be reconstructed' (1970, p. 690).

Evidence of climatic oscillations from terrestrial sources, particularly in temperate latitudes is not sufficiently well defined to permit close correlations with the sea-level oscillations established in north-west England and elsewhere. Both biostratigraphic and lithostratigraphic data bear witness to major climatic shifts, particularly during the Late Devensian stage, but minor shifts inferred from such data do not permit the same sort of precision during the Flandrian stage. Manley (1965) has suggested that during Flandrian II, the mean annual temperature for central England was close to 11.0°C, which is not greatly beyond the range of the contemporary climatic régime. Furthermore, the slight climatic shifts of the Flandrian stage in temperate latitudes are unlikely fundamentally to have affected the composition and structure of the major plant communities, thereby removing the potential for the clear registration and magnitude of climatic oscillations. Rather, the palaeobotanic record contains evidence of succession in response to eutrophication or oligotrophication of the habitat, the latter undoubtedly accelerated by changing climatic parameters such as the increase in precipitation and humidity, the change in the P/E ratio. Such changes have elicited responses in peat stratigraphy, to which the marked recurrence surfaces of raised bogs bear witness. Some recurrence sufaces in Britain together with the construction of trackways occasioned by rising ground water levels and flooding are related to climatic oscillations. Godwin (1966) has indicated that there was an increase in wetness between 2959 and 2575 years BP and has given an average date of 2604 years BP based on thirteen variates for the construction of trackways in England and Wales. On the Lancashire coast, at Pilling, Kate's Pad is dated to 2760±120 (Q.68 Godwin and Willis 1960), and records the age of the trackway prior to renewed peat growth. This date coincides somewhat anomalously with a period of relatively high sea-level, Lytham VIII, and it is probable that the length of this transgression was extended because of sustained storminess in middle latitudes that maintained a sea inlet into Lytham Hall Park. The age of the Sweet Track in Somerset (Coles *et al.* 1973) has been closely dated to 5218±75 (Q.963) and 4887±90 (Q.991) which is the time of a relative fall of sea-level during Lytham VI; the progressive vegetational succession from aquatics to mire plants beneath the Sweet Track is undoubtedly a response to an homogenous fall of sea-level affecting the freshwater table on the Somerset Levels.

The presence of recurrence surfaces in raised bogs throughout North-west Europe and elsewhere has been used as a palaeoclimatic indicator. A decrease in humification and renewed growth of *Sphagnum* peat is interpreted as an expression of increased wetness and an increase of humification resulting from increased dryness. If the time of recurrence or retardation could be shown to be synchronous over wide areas, then inferences could be made on climatic change and other global phenomena such as sea-level movements.

Godwin (1954, 1966) has reviewed the literature on recurrence surfaces and pointed to the complex relationship between renewed peat growth and climatic change. The apparent lack of synchroneity of surfaces within raised bogs and between raised bogs in different areas made correlation very tentative and climatic inductions hazardous. Dickinson, however, (1975) has concluded that 'although there are definite times when recurrence surfaces tend to be formed, the whole series is not necessarily present in any one raised bog' (p. 932). This conclusion is sustained by comparising the ages (converted from T.C-14 to T.corr. using the calibration table of Damon *et al.* 1973b) of recurrence surfaces in Rusland Moss, South Cumbria (Dickinson 1975) and in Draved Mose, Denmark (Aaby 1974, 1976).

	Rusland		Draved
A.D.	1148	A.D.	980
	605		
	427		
	427		
	452		480
			210
B.C.	58	B.C.	
	462		330
	580		620
	914		1420
			3110
			3450

The conversion from radiocarbon to solar years involves standard errors on the converted dates that range from 22 to 125 years, and if these are taken into account both corrected dates fall within the limits of one standard deviation more frequently. Nevertheless, some corrected dates fall outside these limits, and demonstrate the lack of synchroneity of all recurrence surfaces. In part, this is a real effect, but in part it may represent sampling errors, so that data are not strictly comparable.

Notwithstanding these difficulties, Aaby (1974, 1976) has demonstrated from Danish raised bogs, cyclic climatic variations with an apparent periodicity of 260 years, by dating levels where the degree of humification decreased. Increasing wetness was also indicated by increased frequencies of two rhizopods, *Assulina sp.* and *Amphitrema flavum*. Aaby suggests that a decrease in humification indicates increasing humidity, which may result from either lower temperatures, higher rainfall or both.

Such an inferred climatic deterioration should occasion a response in sea-level, and for some of the low sea-level points (taken from Fig. 38 and the age corrected using the calibration table of Damon *et al.* 1973b) there does appear to be a close relationship:

Low Sea-Level Points in West Lancashire	Cool/Wet Points in Danish Bogs	Cool/Wet Points in Rusland Moss
A.D. 630		A.D. 605
432 B.C.	330 B.C.	462 B.C.
1501 B.C.	1420 B.C.	
2964 B.C.	3110 B.C.	
3556 B.C.	3450 B.C.	

There are, in addition, times when the sea-level data and the raised moss data are out of phase. For example, a period of relatively high sea-level is inferred from sand-dune palaeosols at 817 BP (T. corr. 802. or A.D. 1198) which is close in age to a recurrence surface on Rusland Moss. This surface, however, may be related to an increase in humidity and water table in the valley of the Rusland Pool consequent upon a rise in sea-level, for this period of high sea-level coincides with the period of medieval warmth (Lamb 1966) and with climatic amelioration inferred from δ (^{18}O) curve from central Greenland (Dansgaard *et al.* 1975).

To identify slight climatic modulations within temperate latitudes a more refined index for terrestrial and limnic palaeoenvironments is required. For precipitation, the iodine content of biogenic sediments provides a sensitive index (Pennington and Lishman 1971), and the presence of deuterium may provide an equally sensitive index for both precipitation and temperature (Schiegl 1972). The use of geomorphological features and inferred changes in the hydrology of drainage basins during the Flandrian stage (Starkel 1966) is probably too crude an index of climatic change, and is not sufficiently sensitive either to record or to preserve a readily dated record of the climatic shifts of this stage. The same conclusions attach to the records of changing water levels in lake basins.

From the inferred record of climatic oscillations in temperate latitudes there is some difficulty in establishing a relationship with sea-level oscillations, the causal evidence for which resides in high latitude and high altitude areas.

Major climatic shifts and periods of ice cap attenuation or thickening do appear to synchronise in both hemispheres (Epstein *et al.* 1970), and to elicit a major response in world sea-level (Fairbridge 1961; Mörner 1969a). Minor climatic shifts are more difficult to demonstrate from the lack of synchronous, unambiguous behaviour of high latitude and high altitude ice caps and glaciers during the Flandrian stage. There is a certain measure of concurrence; moreover, when such concurrence occurs, a tentative correlation can be made with the succession of marine transgressions and regressions in north-west England.

The rapid rise of sea-level recorded in north-west England from 9270 until 8575 years BP, constituting Lytham I, is undoubtedly a final response in world sea-level to the culminating stages of downwasting of the Scandinavian Ice Sheet, and the attenuation of the Laurentide Ice Sheet. The fall in sea-level and subsequent still-stand or slight rise, is closely related to the Cockburn Glacial Phase of the Laurentide Ice Sheet described by Bryson *et al.* (1969) occurring between 9000 and 8000 years BP. The maximum of the Cockburn re-advance is estimated at between 8500 and 8000 years BP. A nadir is attained on the sea-level curve from north-west England at 8390 years BP, which is close to the time of the minimum sea-level in the Forth valley (Sissons and Brooks 1971). Lytham III comprises a very spectacular rise in sea-level, which is undoubtedly closely related to the disintegration of the Laurentide Ice Sheet (Bryson *et al.* 1969) consequent upon the inundation of Hudson Bay, which became ice free shortly after 8000 BP. The first response in north-west England is the inundation of biogenic sediments by a rising sea-level at 7800 years BP. Difficulties of correlation arise because of the different materials used for dating these episodes: the sea-level curve from north-west England is based on radiometric assays on biogenic sediments, whereas the stages of the disintegration of the Laurentide Ice Sheet derive exclusively from assays on shells (Andrews and Ives 1972).

The transgressive and regressive phases of subsequent transgressions can be related to advance and retreat stages of groups of valley glaciers; the contribution of each in terms of world sea-level movements is difficult to calculate but is unlikely to be very great. Furthermore, the contribution to world sea-level movements and the behaviour of Antarctica in terms of its ice budget over the last 7000 years is imperfectly known. The best that can be achieved at present is to indicate when there appears to be a close correlation between sea-level oscillations and ice advance and retreat stages.

Continued wasting of the remnants of the Laurentide Ice Sheet during Flandrian II probably contributed to a rise in sea-level registered by a transgressive phase in north-west England. Lytham IV occurred between 6710–6157; part of the rise may have been contributed by the deglaciation of the Foxe Basin between 7000 and 6700 years BP (Bryson *et al.* 1969). For the regression, there is some evidence from both hemispheres of ice advance and ice cap thickening. In Banff National Park, Porter and Denton (1967) record a soil beneath ablation till dated at 6020±90 (S.191) and in Antarctica, Denton *et al.* (1971) describe the thickening of the Ross Ice Shelf and give dates of 6100 years BP (Y.2401) and 5900±140 years BP (L.462C) from the Koettlitz Glacier in front of the terminus of the Hobbs Glacier.

There is no evidence for a period of ice retreat towards the end of Flandrian II, and no independent data to support the transgression (Lytham VI) that was under way in 5570 years BP. The evidence is abundant for a climatic deterioration in temperate latitudes and an ice advance stage in high latitude and high altitude areas between 5000 and 4800 years BP, whilst an extensive removal of marine conditions consequent upon a fall in sea-level is recorded in north-west England. This period is particularly critical, for it witnesses the decline in the fortunes of both *Ulmus* and *Tilia*, both of which serve in identifying the boundary between chronozones Flandrian II and III.

Bray (1970) records a global glacial phase between 4500 and 5000 years BP with the maximum ice advance culminating between 4600 and 4800 BP. Goldthwaite (1966) infers ice advance in Alaska from the creation of an ice-dammed lake about 4650 years BP and dates on wood of 4755±180 and also 4680±160 years BP (I.89, Y.9) indicate maximum limits for an early Neoglacial Ice advance in front of the Reid and Carroll Glaciers in Glacier Bay. Porter and Denton (1967) have dated charcoal from a palaeosol overlain by till at Crow's Nest, British Columbia, to 4770±120 years BP (I(GSC).182), and Mercer (1967) has summarised data for Neoglaciation advance stages from both hemispheres. On Mount Garibaldi in British Columbia, ice appears to have overridden tree stumps at some stage shortly after 5260±200 BP (Y. 140, bis), the date on wood from these stumps. Further south in Washington the South Cascade Glacier sheared off trees, the age of wood from which is given as 4700±300 (W.1030). Miller (1969) gives an age of 4960±90 (UW.99) for the maximum limit of South Cascade Glacier.

In the Alps, trunks of *Pinus cembra* embedded in the end moraine of the Oberaar Glacier have been dated to 4600±80 years BP (B.254), giving a maximum age for the ice advance (Mayr 1964). A synchronous advance of the Cadlimo Glacier in Val Piora some 40 km east of the Oberaar Glacier has been inferred by Zoller (1960) on the basis of the recession of spruce forest and expansion of heathland dated at 5150 years BP. This indirect evidence of ice advance and retreat, defined as the Piora oscillation, between 5350 and 4950 years BP based on pollen analytical evidence has been questioned by Mörner (1969a) but supported by Frenzel (1966).

In South America, South Patagonia, Argentina, there is indirect evidence of ice advance from wood dated to 4950±115 buried by lacustrine sediments of a pro-glacial lake formed by the advance of the San Lorenzo Este glacier (Mercer 1967).

A period of relatively low sea-level between 4800 and 4500, and a subsequent rise (Lytham VIa) is recorded in north-west England. Dansgaard *et al.*'s (1969) curve implies a period of amelioration about 4500 years BP, and elsewhere in Europe, a marine transgression is under way (e.g. Calais IV in the Netherlands) supporting the evidence from north-west England. Evidence from Antarctica (Denton *et al.* 1971), however, indicates a period of ice expansion for which there is a date of 4450 years BP from Marble Point (I. 627). This advance is supported by data from South America, where in the Sierra de Sangra in southern Patagonia, a minimum date for the culmination of maximum advance of the Narvaez Glacier is given by Mercer (1967) as 4320±110 years BP. If both hemispheres were out of phase during this period, and if a rise of sea-level was under way with a maximum at about 4500 years BP, this proportion of the eustatic potential must have been realised by the withdrawal of glaciers in the northern hemisphere only.

Lytham VII is well supported. In South America, the San Rafael Glacier in the southern Andes reached its post-glacial maximum before 3740±400 BP from which position it had withdrawn at this time, and withdrawal of the Moreno Glacier in southern Patagonia had also taken place before 3830±115. In the Yukon, Rampton (1970) has dated wood beneath till to 3280±130 and 3300±130 years BP (GSC.933 and GSC.1003). From the Gooschenenalp Series in Switzerland, there is a date of 3340±120 (B.382) on *Larix* wood, overridden by ice at some stage after this date (Gfeller and Oeschger 1963). If this indicates a response to a global climatic deterioration, then it can be correlated with the fall of sea-level and the culminating stages of Lytham VII dated to 3150 years BP.

This oscillation was short-lived, for Lytham VIII is dated at 3090 to 2270. The transgression stage is poorly supported, but the regression stage is heralded by a global glacial phase, recognised by Bray (1970) as the Sub-Atlantic Minimum, culminating between 2600 and 2800 years BP. Dansgaard *et al.*'s (1969) $\delta(^{18}O)$ curve shows low values between 2100 and 2500 years BP. In Alaska, ice advanced down Muir Inlet, and wood embedded in outwash gravels is dated at 2340 years BP (I.1612) (Goldthwaite 1966). Furthermore, tree stumps embedded in Glacier Bay till are dated at 2120 years (I.1610).

Fulton (1971) argues for extensive glacier advances throughout the Yukon and Alaska between 3000 and 2500 years BP. He cites the example of the Tiedemann Glacier on the eastern flank of Mt. Waddington which advanced over a bog, for which there is a date of 2940±130 (GSC.938), withdrawing from the outermost moraine by 2250±130 BP (GSC.948).

The Chelon Glacier above the Goeschenalp reservoir in Switzerland advanced after 2280±120 (B.380) (Porter and Denton 1967). In South America, the outlet of the South Patagonian icefield, the Uppsala Glacier, formed about 2100 years BP (Mercer 1967).

It would appear that even while the transgression was under way in north-west England a global climatic deterioration had set in, and either the response in global sea-level adjustment was delayed or the length of time Lytham VIII was recorded in north-west England was the result of local site conditions. It is suggested that marine sedimentation was sustained in Lytham Hall Park because a period of storminess maintained the area as an inlet open to the sea.

Lytham IX is supported only by evidence from the Cape Century Ice Core (Dansgaard *et al.* 1969), by Wiseman's carbonate curve (1966) and by Wollin *et al.*'s (1971) foraminiferal curves. The evidence of ice advance and retreat phases tends to run contrary to these trends, although deriving only from the Yukon (Rampton 1970) and the Alps (Mayr 1964). Probably a maximum date of 1520±130 (GSC.751) on wood subjacent to a 25 foot till unit deposited by the Klutlan Glacier in the Yukon has been obtained, although in Europe Mayr has established a period of ice advance to 1550-1220 years BP. Denton and Karlén (1973) and Karlén and Denton (1976) have identified five periods of glacier advance in the Kebnekaise Mountains of northern Sweden during the Flandrian stage that are supported by evidence from the Sarek National

Park, also in northern Sweden, and the St. Elias Mountains in southern Yukon Territory and Alaska. These periods of glacier expansion synchronising in two continents bear witness to significant climatic shifts in the northern hemisphere during the Flandrian stage and should be associated with sympathetic sea-level movements. The periods of glacier expansion are dated to 7300–7500 BP, 5300 BP, 4300–4600 BP, 2200–2800 BP and 1000–1350 BP. In north-west England periods of low sea-level are identified from Fig. 38 at 7120 BP between transgressions Lytham III and IV, at 5282 BP within the complex transgression Lytham VI, at 4282 BP between Lytham VI and VIa, at 2284 BP between Lytham VIII and IX and at 1320 between Lytham IX and X.

Many problems attach to the dating of glacier advance and retreat stages, and the exact state of the total mass budgets of glaciers and ice sheets is difficult to establish. Furthermore, the relaxation times of climatic change, change in total mass budget, and sea-level response are not known adequately. However, the evidence presented here does indicate a tendency towards synchronization of global events, however crude the index points are, both for dating and for inferring global events. There does appear to be a convergence of data, derived from different sources to support the oscillating sea-level curve closely related to climatic shifts, and the concurrence of these two globally related phenomena substantiates the arguments and data of Fairbridge (1961) and Mörner (1969a). Furthermore, it appears that low-lying coastal zones, being sensitive to changes in world sea-level of both high and low amplitudes, register climatic oscillations. This conclusion strongly supports Kukla's (1969) inferential argument of a record of climatic change during the Flandrian stage, based on eustatic movements of sea-level recorded in suitable coastal areas.

CHAPTER 6

Conclusions

Flandrian Marine Sequences

In north-west England, ten distinct marine transgressions have been proved during a period spanning all but the first millennium of the Flandrian stage. The transgressions are registered as clays, silts and sands, laid down under brackish-water or full marine conditions, and are separated by biogenic sediments of limnic, telmatic or terrestrial origin. Radiocarbon assays on biogenic material intercalating the marine sediments permit the establishment of time limits for the transgressions identified. The most complete marine sequence for the Flandrian stage is recorded from the lagoonal and tidal-flat zones of Lytham, beneath an over-burden of dune sands or beneath recently reclaimed estuarine marshes. The Flandrian marine transgressions identified in north-west England are given the name of Lytham after the type area (see above, p. 113). This transgression sequence is compared with sequences established in North Wales, north-east England, the Fenlands, the Somerset Levels, and the comparison extended to the continent. A tentative pan-North-west European correlation is proposed, and the time limits of all the transgressions identified shown graphically on Fig. 40. The beginning and end of many transgressions in North-west Europe are synchronous, and where transgressions appear diachronous their time limits have been advanced or retarded either by the index points selected or by regional factors, such as subsidence or uplift. There is a need for many more absolute dates on the transgressive and regressive phases of the transgressions recognised in all areas, so that a strong correlation can be established. The establishment of synchronous marine episodes throughout North-west Europe, irrespective of local and regional factors, would permit tentative conclusions about the amplitude of eustatic and palaeoclimatic oscillations.

The effect of Flandrian sea-level changes on stratigraphic successions

The maximum thickness of sediments, which accumulated within the tidal flat or lagoonal zones under full marine or estuarine conditions is 25 metres. This thickness is recorded in Morecambe Bay, where the shape of the Bay and the tidal streams allowed little opportunity for quiet water, biogenic sedimentation during the rapid rise of sea-level in Flandrian I. Elsewhere in North Wales and north-west England, Flandrian marine and brackish-water sediments reach maximum thicknesses of 17 metres at Foryd and 14 metres at Lytham Common. Greater thicknesses have been recorded in the basins of the Irish Sea and the Celtic Sea, adjacent to which, in Cork Harbour, Stillman (1968) has recorded 27 metres of Late Devensian and Flandrian marine sediments overlying a freshwater biogenic sediment at −55 metres I.O.D.

It is only adjacent to, and landward of, the present coast that intercalated biogenic sediments of brackish and freshwater, telmatic or terrestrial origin allow subdivision of the stratigraphic column into radiocarbon-dated time units. The following subdivision within the lagoonal and tidal flat zone of north-west England is recognised:

1. *Lower peat*. A hard, dry fissile peat is recorded at altitudes from -17.0 to −9.0 metres O.D. The peat comprises either a woody detritus or a freshwater gyttja. Microfossil analyses indicate approaching marine conditions. The lower peat can be assigned to chronozones Flandrian Ib, c and d. The lower peat is occasionally underlaid by marine sediments on the weathered surface of the Devensian till. In every case it is overlaid by marine facies. The lower peat is encountered beneath the present coast or immediately adjacent to it, at depths from 1920 cm to 1113 cm from the present surface, and with the overlying marine sediment can be assigned to the *Assise de Calais* (Dubois 1924).

2. *Marine sediments of Lytham I-III age*. These sediments, which accumulated between 9200 and 7200 years BP, comprise silts and sands with a faunal assemblage characterised by *Cerastoderma*, *Mytilus* and *Turritella*. They attain maximum thickness of all the Flandrian marine sediments from lagoonal and tidal flat situations, and reflect a rapidly rising sea-level during Flandrian I. Sediments referable to this period attain a maximum thickness of 15 metres. The formation can be subdivided on the basis of thin intercalated biogenic sediments, reflecting oscillations in sea-level during the general rise. These biogenic sediments, like the lower peats, are dry, compact and fissile, but are usually dominated by mono-cotyledonous macrofossils, and may be either gyttjas or herbaceous, detrital peats. The proximity of marine conditions during their formation is indicated from the microfossil assemblages. Absolute dates on these biogenic sediments permitted the subdivision of the formation into Lytham I, II and III, all of which constitute transgressions landward of the present coast. The formation extends from −1.0 to −16.0 metres O.D., and together with the lower peat constitutes the *Assise de Calais*. It is homologous to the Old Tidal Flat or Calais Deposits recognised in the Netherlands (Jelgersma 1961; Hageman 1969), except that there the altitude limits are −4.0 to −18.5 metres N.A.P., and the time limits are 7950 to 3750 years BP.

3. *Marine Sediments of Lytham IV-IX age*. These sediments are characteristically silty clays, often very rich in biogenic partings, such as the rhizomes of *Phragmites*, and with beds of *Scrobicularia*. They are estuarine or brackish water lagoonal sediments, which formed after the main restoration of sea-level was completed, but during a period in Flandrian II and III when sea-level was rising slowly and was characterised by many oscillations. During periods of relatively high sea-level, transgressions occurred, and during periods of relative low sea-level, biogenic sedi-mentation occurred on the regressive surfaces of the transgressions. The

transgression facies are characteristically thin and rarely exceed two metres. The maximum thickness of the formation in the type area is 456 cm comprising both biogenic sediments and transgression facies. The biogenic sediments can be of limnic, telmatic or terrestrial origin, and comprise gyttjas, monocotyledonous and woody detrital peats and bryophyte and monocotyledonous turfas. A fundamental subdivision of this formation can be made on the basis of the spatial extent of the transgressions. Lytham IV to VI are particularly extensive, and Lytham VI throughout north-west England extended landward to as great an extent as some of the earlier transgressions comprising Lytham I-III. Subsequent transgressions, Lytham VII-IX, were confined to small embayments adjacent to estuaries; their confinement resulted from the considerably reduced rate of sea-level rise, the rate of biogenic accumulation, the development of coastal sand-dune systems consequent upon the slight sea-level fluctuations during this period, and the removal of areas potentially suitable for inundation through the operation of the last two factors. Because this formation was a product of a relative sea-level close to present Mean Sea Level, it can be correlated with the *Assise de Dunkerque*, which was characterised by Dubois by the contemporary faunal assemblages dominated by *Cardium* and *Scrobicularia*. In the Netherlands, the heights of the Young Tidal Flat or Dunkirk Deposits is similar, −4.0 to +1.2 metres N.A.P., but the time limits are restricted to 3450 to 1150 years BP.

Isostatic Recovery

The coast of north-west England is critical to a consideration of delevelling during the Late Devensian and Flandrian stages. The Lancashire coast occupied a position marginal to the area of maximum ice loading, and it is likely to have suffered both positive and negative movements during these stages.

The heights of the transgressive and regressive surfaces of marine transgressions recorded in North Wales, Cheshire, south-west Lancashire and the Fylde of Lancashire show little distortion over a distance of 45 km when projected onto a line orthogonal to the supposed ice front. Further north, particularly north of the southern margin of Morecambe Bay, there is a sudden increase in the gradients of the transgressive surfaces: the conclusion that a hingeline occurs across Morecambe Bay, separating areas to the north where ice loading was greater than areas to the south, is unavoidable. The change in altitude of Lytham II across Morecambe Bay may reinforce this conclusion; it may provide data for the behaviour of the upper mantle during the application and removal of crustal loads, or it may reflect tectonic activity along the margins of the East Irish Sea Basin. The resolution of this enigma resides, in part, in the establishment of a Late Devensian glaciation chronology along the eastern margin of the Irish Sea. The stages of deglaciation are imperfectly known. The ice limit suggested by Mitchell (1972), for 18,000 years BP as a lobe extending from the Lake District to the Cheshire Plain, is unlikely, and the well-marked terminal features in the Cheshire Plain (Shaw 1972) are probably of greater antiquity. Mitchell neglects to consider the position and age of the Kirkham end moraine, which bisects the Fylde and has been described by Gresswell (1967). Furthermore, the Fylde was ice free well before 12,320±155 years BP, which is the date on a basal gyttja from a kettle on the foreshore at Rossall (Tooley 1977a). It is suggested that 18,000 years BP the ice front occupied the line of latitude from the northern part of the Isle of Man to the Cumbria coast. This suggestion requires corroborative field evidence, but is sympathetic to the hypothesis that since Flandrian Ib uplift in the Fylde and south-west Lancashire has been negligible and has permitted the registration of a successive series of marine transgressions there. This argument (Tooley 1969) was reinforced by the near-horizontal and sub-parallel condition of the transgression surfaces around Liverpool Bay and to the north and west of the Bay.

The nature of the evidence from Wales and Somerset, however, throws doubt on its validity, for there are systematic altitudinal discrepancies between the transgression surfaces in South Wales and Somerset, and in Lancashire. Furthermore, the altitudinal differences decrease for each transgression, except Lytham I, during the Flandrian stage, thereby strengthening the case for modest uplift during Flandrian I and II within the type area, but at no time of such magnitude that each oscillation of sea-level could not be registered directly in west Lancashire. The following altitudinal differences are recorded between transgression surfaces in the type area and their homologues in Wales and south-west England:

Lytham I	2.85 m
Lytham II	9.56 m
Lytham III	6.81 m
Lytham IV (Transgressive)	4.92 m
Lytham IV (Regressive)	4.26 m
Lytham V	2.05 m
Lytham VII	1.07 m

If the height differences for Lytham I-IV are deducted from the mean sea-level curve from west Lancashire displayed in Fig. 38, then the curve coincides with Mörner's eustatic curve displayed in Fig. 42 and thereafter, without further adjustment, both curves occupy almost identical altitudinal positions. This would imply that, if Mörner's curve approaches very closely to the eustatic curve, global sea-level changes are registered directly in south-west Lancashire and the Fylde from 5500 years BP to the present.

This degree of coincidence, remarkable as it is, may very well be illusory, and three important conditions should be considered when examining the data from western Britain. Firstly, no allowance has been made for the considerable tidal range or for differences in altitude in relation to Ordnance Datum, at which M.H.W.S.T. intersects different points on the coast or for changes in tidal range resulting from the changing geometry of bays and inlets during the Flandrian stage (See Jardine 1975). Secondly, the sample population is very small, and altitudinal differences may represent random errors. Thirdly, sites from Wales adjacent to the Bristol Channel, have not been related to their contemporary sea-levels satisfactorily, and significant errors are implicit and are attendant upon such comparisons.

The following conclusions, deriving from a consideration of the evidence for isostatic recovery in north-west England, can be made with a little more confidence:

1. From the increase in gradient and inflexion of the transgression surfaces across Morecambe Bay, this area is regarded as a hinge-line, and has operated as such throughout the Flandrian stage.

2. The outer limit of Flandrian raised beaches is the southern margin of Morecambe Bay. Wright's zero isobase must be shifted northwards by 125 km to bisect the north-west coast of England in the vicinity of Heysham, and the north-east coast of England in the vicinity of Alnmouth. The evidence supports neither the argument, advanced by Whittow (1971), to shift the zero isobase south to the latitude of Borth, nor the approximate outer limits of both late- and post-glacial shorelines shown bisecting the western coast of Britain in Furness and the Lleyn Peninsula respectively, by Stephens and Synge (1966). There is no unequivocal evidence for Late Devensian raised beaches in Lancashire and Cumbria, and Sissons (1967) has already noted their absence in Cumbria (see discussion in Tooley 1977b). The evidence undoubtedly resides in the Irish Sea basin and has been buried by Flandrian marine sediments, or eroded during the restoration of sea-level.

Relative Sea-level Changes

The evidence from north-west England is for an oscillating sea-level during the Flandrian stage, accomplishing the final restoration of sea-level. The following oscillations in *relative* sea-level have been identified in west Lancashire (Fig. 38):

1. Relative sea-level rose rapidly from the end of Flandrian Ib until the middle of Flandrian Ic, at a rate of about 1.8 cm/year, to a maximum altitude of -14.0 m O.D., from which altitude it then fell to one of -15.2 m O.D.

2. During the latter half of Flandrian Ic sea-level rose by a metre or was stationary between 8390 and 7995 years BP.

3. Shortly after 7800 years BP sea-level rose extremely rapidly to an altitude of -6.5 m O.D.

4. After 7600 years BP the rate of rise slackened to 0.3 cm/year until 6980 years BP. During this period and the preceding one, the most extensive penetration by the sea occurred landward of the present coast.

5. Sea-level fell after 6980 years BP, to about -4.3 metres O.D., and there was an extensive removal of marine conditions in south-west Lancashire; on Downholland Moss, the removal of marine conditions extended over a surface greater than 2 km in width.

6. Sea-level rose between 6800 and 6300 years BP by over a metre and reached an altitude of -3.2 m O.D.; during this general rise there is some evidence for a slight oscillation. The Lytham-Skippool valley was inundated and the east and west parts of the Fylde were separated.

7. About 6000 years BP sea-level fell, and there is evidence for an extensive removal of marine conditions once again throughout south-west Lancashire and the Fylde.

8. Sea-level rose rapidly between 6000 and 5800 years BP at a rate approaching 0.6 cm/year, and a maximum altitude of -2.2 m O.D. was attained.

9. Sea-level fell between 5775 and 5500 years BP to a nadir of about −3.4 m O.D.

10. Between 5500 and 5000 years BP, sea-level rose at a rate of 0.5 m/year.

11. The opening of Flandrian III, from 5000 to 4897 years BP, is marked by a rapid fall in sea-level from −1.1 to −1.9 m O.D., and a very extensive removal of marine conditions throughout north-west England. The marine influence was removed finally from the Lytham–Skippool valley, and after two thousand years of separation, the east and west Fylde became a continuous land area.

12. Sea-level rose to a maximum altitude of −1.0 m O.D. about 4500 years BP and subsequently fell to an altitude for which there is as yet no index point.

13. A rise of sea-level was under way by 3700 years BP, and at about 3400 years BP probably rose slightly above present Ordnance Datum on the Fylde coast.

14. Sea-level fell subsequently to −0.7 m O.D. about 3100 years BP.

15. Sea-level rose and transgressed biogenic sediments 3090 years BP, at an altitude of −0.5 m O.D., and continued to rise, exceeding present sea-level, to an estimated altitude of +1.0 m O.D.

16. Sea-level fell to a minimum value close to present Ordnance Datum at 2270 years BP.

17. Sea-level rose to a maximum altitude of +1.2m O.D. c. 1800 years BP.

18. Sea-level fell during the succeeding 500 years, to a minimum altitude of +0.4 m O.D.

19. Subsequently, sea-level rose and fell, reaching maximum altitudinal values between 830 and 805 years BP.

This curve shows the oscillations of Mean High Water Mark of Spring Tides, and the data transformed, by deducting 4.2 metres, to show oscillations of Mean Tide Level which is 0.22 metres above Mean Sea Level at St. Annes. This curve has been plotted on Fig. 42, for comparison with the sea-level curves of Fairbridge (1961), Jelgersma (1961, 1966), Shepard (1963), Mörner (1969a), Ters (1973) and Kidson and Heyworth (1973).

Between 5500 years BP and the present, the Mörner curve and the curve from west Lancashire are more or less coincidental. The altitudes of both curves are also similar during Flandrian Ib. However, the intervening period from 9000 to 5500 years BP shows a similar pattern but separated by some 5 metres, the difference becoming progressively less towards the end of Flandrian II. This difference is reduced by deducting the estimated amount of isostatic recovery that has affected the index points in north-west England during this period. Both curves then become concurrent for most of the Flandrian stage.

The curves displayed in Fig. 42 emphasise the dichotomy between different sea-level studies: one group argues for an oscillating sea-level (Fairbridge 1958, 1961; Mörner 1969a; Ters 1973), the other for a continuously rising sea-level (Jelgersma 1961, 1966; Shepard 1967; Kidson and Heyworth 1973). Mitchell (1977) subscribes to the conclusions of the latter group even though his suggestive sea-level curve shows both long and short period oscillations.

The data presented here provide unequivocal evidence for an oscillating sea-level during the Flandrian stage. The magnitude of each oscillation is open to debate, and in part reflects local conditions. There is a remarkable similarity between the temporal pattern of the sea-level oscillations and eustatic changes, which is not the result of the operation of random factors. The sea-level curve from north-west England supports the oscillating theory of sea-level, propounded by Fairbridge, and supported by Mörner and Ters, and is in controversion to the continuously rising sea-level theory.

Finally, the sea-level curve from north-west England has some important regional implications.

The position of relative sea-level 9000 years BP was about −21.0 m O.D. and by 8600 years BP had risen to −14.0 m O.D. During this period, a proto-Morecambe Bay was in existence, and the present coasts of North Wales, south-west Lancashire and the Fylde had been transgressed at altitudes ranging from −12.3 to −9.6 m O.D. The conclusion is inescapable that the Irish Sea had come into formation by the first millennium of the Flandrian stage, and the land-bridge between England and Ireland had been severed before this time, and some 1.5 to 2 thousand years before the closure of the North Sea land-bridge. It is doubtful whether the Irish Sea Basin was available for plant and animal migration during the first millennium of the Flandrian stage, as Mitchell suggests (1960, 1963, 1972), and the proposed closure throughout the Flandrian stage has important implications for the biogeography of

Ireland and the Isle of Man. Thomas (1977) has proposed that there was a marine accompaniment to the decay of Devensian ice in the Irish Sea, and there is a possibility of a proto-Irish Sea in existence, at least in part, during the Late Devensian to which glacio-marine sediments around the East Irish Sea Basin may bear witness (Tooley 1977b).

The sea-level curve shows a rapidly rising sea-level relative to the land in north-west England between 9200 and 6800 years BP, during which extensive marine transgressions occurred, and there was little opportunity for biogenic sedimentation. In the subsequent six millennia, sea-level continued to rise, at a greatly reduced rate, and each oscillation in sea-level was registered by a marine transgression of progressively limited extent. On three occasions during the last 3200 years, sea-level rose above its present level, attaining maxima about 2650, 1800 and 800 years BP. From this level, sea-level appears to have fallen relative to the land, notwithstanding the slight downwarping recorded by Valentin (1953).

The conclusion from these data is that the Lancashire coast is becoming less susceptible to marine transgressions, and if sea-level maintains its general decline in this area, then low-lying areas within the lagoonal, tidal flat and perimarine zones will become progressively less prone to marine inundations, and the chances of the return period marine flood on the scale of the 1720, 1833, 1907 and 1927 occurrences (Barron 1938; Beck 1954; Blundell 1720) will become considerably reduced.

There are also national and international implications. It has been demonstrated in Chapters 4 and 5 that the restoration of sea-level during the Flandrian stage was achieved by a series of oscillations closely associated with climatic oscillations. Some of the higher sea-level stands were achieved very rapidly: for example, Lytham III has been established from two index points (Table 1, No's 48 and 57) which demonstrate that sea-level rose over 7 metres in about 200 years. If the standard errors on the radiocarbon dates are taken into account, then the rise of sea-level could have been achieved in a period as short as 70 years or as long as 360 years. This surge of sea-level was the consequence of the break-up of the Laurentide Ice Sheet over Hudson Bay. Mercer (1978) has noted the inherent instability of the West Antarctic Ice Cap, buttressed by the Ross and the Ronne Ice Shelves, and a rise of 5 metres in sea-level consequent upon its melting. The effect on low-lying coastal areas in the developed and developing world of such a rise given the existing tidal ranges would be disastrous. Submergence of much of Florida, the Netherlands, the coastal hinterland of the Indian sub-continent, i.a. would follow. Port and harbour installations, industrial and residential areas, transport facilities and energy generating installations would all be affected either directly or indirectly. In the context of the United Kingdom, areas of high risk are south-east England, the Fenlands and low-lying areas flanking the Humber, Tees, Ribble, Mersey and Severn estuaries. In all these areas the rate of development below + 7 m O.D. has intensified during the last twenty-five years. It is interesting to note that all but one of the Central Electricity Generating Board's Nuclear power stations are located on the coast at altitudes ranging from +4 to +12 m O.D. and given a surge of sea-level, comparable to Lytham III, at high tide the sites would be standing up to 12 metres in sea-water. In 1970, it was recommended (Tooley 1971) that there was a need to measure long-term trends in the magnitude and direction of relative and absolute changes in sea-level, and to identify high risk coastal zones as a basis for rational long-term planning of these areas: this need still requires to be met.

Sea-level changes and the geoid

It is well known that the earth is not spherical, but flattened at the poles and bulging at the equator: the earth approximates to a pear-shape in profile with sea-level at the North Pole 44.7 m further from the equator than at the South Pole (King-Hele 1967 1975).

It has been assumed that the sea surface parallels the shape of the earth, but recently geoid solutions have demonstrated the existence of high and low values distributed over the globe. The Goddard Earth Model 6 geoid map (in King-Hele 1975) shows depressions of the ocean surface of -112 m south of the Indian sub-continent and high values of +78 m near New Guinea. The irregularities in the free ocean surface (the geoid) are caused by the earth's gravity, itself affected by the structure and density of the earth, and by its rotation (Mörner 1976b).

Mörner (1976b) has drawn attention to the fact that the geoid pattern — the distribution and amplitude of high and low values — is not stable in geological time. Changes in geoid configuration will result from the redistribution of mass occasioned by the repetitive series of glaciations and deglaciations during the Quaternary and concomitant changes in sea-level (Jensen 1972), which will affect the earth's rotation. The implications for sea-level studies are apparent: the free ocean surface can intersect different continents simultaneously at different absolute altitudes (the contemporary range in altitudes of the geoid shown on GEM 6 is 180 metres which is close to the 'measured' difference between low sea-levels of the last glacial stage and high sea-levels of this interglacial stage) and the changing pattern of the geoid may occasion a record of a marine transgression or regression, when no addition or subtraction of water to the ocean basins has occurred.

The establishment of a precise geoid relief map, and the probability that the pattern of relief has changed does not invalidate the attempt to establish relative sea-level curves for small homogeneous areas. It does, however, add a further imponderable to the problem of regional, continental and inter-continental correlation, and renders the prime aim of the International Geological Correlation Programme on Holocene Sea-Level Movements to establish a single eustatic sea-level curve open to question, as Mörner (1976b) observed so perspicaciously.

APPENDIX

SAAR, A. du. (1969) Diatom Investigation of a Sediment Core. Downholland Moss-15

Geological Survey of the Netherlands. Department of Diatoms and Ostracods. Report No. 150*

Introduction

In order to trace changes in sedimentation during the deposition of the Downholland Moss profile, taken in the low coastal area near Formby, 10 sediment samples have been examined for their diatoms (Appendix, Table 1, and Fig. 1).

For the diatom analysis of the samples, the qualitative method was used. From each sample, one diatom mount has been examined, and the species identified have been marked on analysis lists. The presence of the species in the samples was estimated and noted on these lists in the following way:

> RR   very rare
> R   rare
> C   common
> A   abundant
> AA   very abundant

For marking species on the lists, different colours have been used to indicate the chloride concentration that is favoured by each species according to the classification of van der Werff (1958-66). The explanation of the colours is as follows:

red	marine	>17000	mg Cl⁻/1
brown	marine-brackish	1000–1700	—
orange	brackish-marine	5000–1000	—
yellow	brackish	1000–5000	—
light green	brackish-fresh	500–1000	—
dark green	fresh-brackish	100– 500	—
blue	fresh	<100	—

For each sample the Marine-Brackish-Fresh (M-B-F) ratio, introduced by van der Werff, has been calculated. This calculation is as follows: all species of the assemblage of a sample that belong to the same chloride group are summed up and the group totals are expressed as percentages of the total number of species in the sample. After that the percentages of the marine, the brackish and the fresh groups are added and these additions form the M-B-F ratio.

This ratio immediately gives an impression of the composition of the diatom assemblage of the sample and shows whether marine, brackish or freshwater conditions predominated during sedimentation.

To show the changes in the chloride content during sedimentation of the Downholland deposits a M-B-F diagram has been drawn (p. 204). In this table an indication of whether a species belongs to the plankton- the epiphytic- (or epontic) or the benthonic flora is given. benthonic flora is given.

*Printed by kind permission of the Director, Rijks Geologische Dienst, Haarlem, The Netherlands.

DOWNHOLLAND MOSS - 15
Distribution of the diatom species

Depth of the samples in cm below top of the core

Distribution of the diatom species	Plankton	Epiphytos	Benthos	55	60	85	100	119	140	165	220	225	265
Marine													
Actinocyclus ehrenbergii Ralfs	P						RR						
Biddulphia aruita (Lyngbye) Brebisson	(P)	E										RR	
rhombus (Ehr.) W. Smith		E				RR	RR	RR		RR		RR	
Coscinodiscus nitidus Gregory	P									RR			
perforatus var. cellulosa Grunow	P						RR	RR		RR		RR	
spec.	P						RR		RR	RR	RR		
Cymatosira belgica Grunow		E						RR					
Diploneis fusca (Greg.) Cleve			B				RR						
suborbicularis (Greg.) Cleve		E							C	C		RR	
Melosira sulcata (Ehr.) Kützing	(P)		B		RR	R/C	A	C/A	C	R/C	C	R/C	R/C
westii W. Smith	(P)		B		RR	RR	R	R	RR	RR	R	RR	R/C
Navicula cancellata Donkin			B							RR			
distans (W. Sm.) Schmidt			B		RR		R	RR		RR		RR	RR
pennata Schmidt			B									RR	
Nitzschia panduriformis Gregory			B				RR				RR	RR	
Opephora pacifica (Grun.) Petit		E				RR							
Pinnularia trevelyana (Donkin) Rabenhorst			B									RR	
Plagiogramma staurophorum (Greg.) Heiberg			B					RR	RR	RR			
Podosira stelliger (Bail.) Mann	(P)	E				R	A	C	R	R	R/C	R/C	R/C
Surirella fastuosa (Ehr.) Kützing			B			RR	RR		RR			RR	RR
Trachineis aspera (Ehr.) Cleve			B					RR					
Triceratium alternans Bailey	(P)		B				RR	RR					
favus Ehrenberg	(P)		B	RR	RR	RR	RR			R			R
Marine-Brackish													
Actinoptychus undulatus (Bail.) Ralfs	P					R	C	R	R	RR	R	RR	RR
Coscinodiscus excentricus Ehrenberg	P						RR	RR			RR	RR	
Diploneis didyma Ehrenberg			B		R/C	AA	C/A	A	AA	AA	C/A	R	RR
smithii (Breb.) Cleve		E					AA		R/C	A	R/C	C	
fo. rhombica Mereschkowsky		E			RR	R			A	R			
stroemii Hustedt		E	(B)			R	RR	R/C			RR		
Navicula flanatica Grunow			B					RR					
humerosa Brebisson			B				RR						
marina Ralfs			B			C		RR	R/C	R	R		RR
Nitzschia acuminata (W. Sm.) Grunow			B						RR				
granulata Grunow			B						RR				
Raphoneis amphiceros Ehrenberg		E			RR	R	C	C	R	RR	R	R	RR
surirella (Ehr.) Grunow		E					R	C	R	RR	R	R	R
Brackish-Marine													
Achnanthes brevipes Agardh		E									RR	RR	
Amphora exigua Gregory			B									R	
Diploneis estuari Hustedt			B							RR			
Nitzschia obtusa W. Smith			B								AA	C	
punctata (W. Sm.) Grunow			B			C	R/C	C	C	C	C	R	R
Scoliopleura tumida (Breb.) Rabenhorst			B		RR	RR	RR	A	R	R	RR	RR	RR
Stauroneis gregory Ralfs			B					RR				RR	

Brackish

Taxon	L	1	2	3	4	5	6	7	8	9	10
Achnantes hauckiana Grunow	E				RR						
Amphora commutata Grunow	B						RR	RR			
proteus Gregory		R		RR	R	RR					
Caloneis amphisbaena var. subsaline (Donk.) Cleve								RR			
formosa (Greg.) Cleve		R	AA	RR	RR	C	C	C		R	R
Campylodiscus echeneis Ehrenberg			R	RR		RR			RR		
Cyclotella striata (Kütz.) Grunow	P	RR	RR				RR	RR	R	R	R
Diploneis interrupta (Kütz.) Cleve	B	R/C	C	R	R	R/C	RR	RR	R	R/C	RR
Gyrosigma speneeri (W. Sm.) Cleve	B				C						
Navicula digitoradiata (Greg.) A. Schmidt	B				RR	R				RR	
elegans W. Smith	B		RR							AA	C/A
peregrina (Ehr.) Kützing	B	R/C	AA	RR	RR	R	C/A	RR		AA	AA
Nitzschia bilobata W. Smith	B					RR	RR				
navicularis (Breb.) Grunow	B	R	A	AA	A	A	A	C/A	R	C	R/C
vitrea Norman	B					RR				R	
Rhopalodia gibberula (Ehr.) O. Müller	E					R				R	
musculus (Kütz.) O. Müller	E				RR		R		R		
Surirella striatula Turpin	B					RR					

Brackish-Fresh

Taxon	L	1	2	3	4	5	6	7	8	9	10
Achnantes brevipes var. intermedia (Kütz.) Cleve	E					RR	RR				
Cyclotella meneghiniana Kützing	P								C/A		
Fragilaria construens var. subsalina Hustedt	E						R		R		
Nitzschia hungarica Grunow	B					RR	RR				
scalaris (Ehr.) W. Smith	B									AA	R/C
sigma (Kütz.) W. Smith	B				RR	C	R/C				

Fresh-Brackish

Taxon	L	1	2	3	4	5	6	7	8	9	10
Amphora ovalis Kützing	B						R/C	R/C		C	C
var. libyca (Ehr.) Cleve	B						RR			C	R
Anomoeoneis sphaerophora (Kütz.) Pfitzner	B	RR							C		
Caloneis bacillum (Grun.) Mereschkowsky	B								RR	RR	
silicula (Ehr.) Cleve	B								R	R	
var. truncatula Grunow	B								RR	RR	
Cymbella aspera (Ehr.) Cleve	E (B)								RR		
spec.	E (B)	RR									
Diploneis ovalis (Hilse) Cleve'	B		RR		R	C			RR	C	
Epithemia turgida (Ehr.) Kützing	E			RR							
Fragilaria construens var. venter (Ehr.) Grunow	(P) E								A		
Gomphonema acuminatum Ehrenberg	E								R/C		
var. brebissonii (Kütz.) Cleve	E								C		
lanceolatum Ehrenberg	E								C		
Hantzschia amphioxys (Ehr.) Grunow	B	RR							RR		
Mastogloia smithii var. lacustris Grunow	B								RR		
Navicula cincta (Ehr.) Kützing	B				RR						
pusilla W. Smith	B						R/C		RR	R/C	
radiosa Kützing	B								R	RR	
viridula Kützing	B									RR	
Pinnularia maior Kützing	B								R	R	
viridis (Nitzsch) Ehrenberg	B								R/C	C	
spec.	B	RR					RR		R	R	
spec.	B								R		
Rhopalodia gibba (Ehr.) O. Müller	E						RR				
Surirella ovata Kützing var.	B								RR		

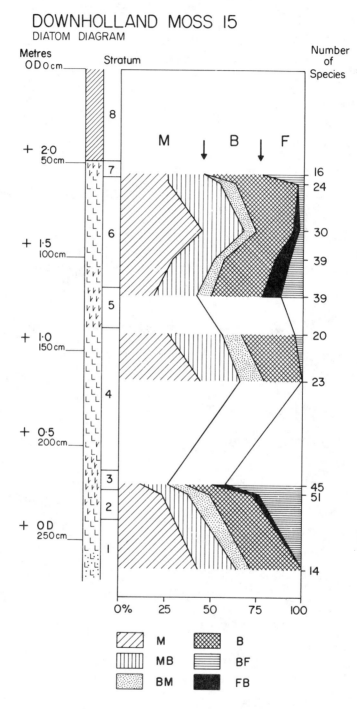

DOWNHOLLAND MOSS 15
DIATOM DIAGRAM

Discussion of the results

The sample at 265 cm taken from the sandy part of the lowermost blue clay layer is poor in diatoms. Only 14 species have been found in but small numbers and mostly as fragments or corroded valves.

In the diatom assemblage, no species that indicate low chloride contents (1000 mg/l) occur. The M-B-F ratio of 65-35-0 indicates no fully marine but brackish to strongly-brackish condition. The character and the state of preservation of the assemblage point to sedimentation in the tidal area.

The sample at 225 cm from the overlying organic clay is rich in diatoms, especially in individuals. The assemblage is composed here of marine, brackish and indifferent freshwater forms. The M-B-F ratio of 37-38-25 shows a clear decrease of the marine influence.

In the assemblage the decrease of the number of marine species goes together with the presence of a rich brackish benthonic flora, which certainly forms the only autochthonous component. This brackish flora and the presence of freshwater species indicate the beginning of freshening of a highly silted up area.

The decrease of the marine influence is still stronger in the *Phragmites* peat sample at 220 cm. At this depth the M-B-F ratio is 26-31-43. For the first time the value of the indifferent freshwater species is higher than the other values.

In the assemblage the brackish benthos group is still well represented but this goes for a number of fresh water species too. This indicates that the freshening of the area had become almost a fact — the state of isolation from the sea is nearly reached. Probably full isolation will be reached somewhat higher up in the peat. As to the sample the only conclusion can be that a brackish marshlike area turned to a slightly brackish reed swamp.

The sample at 165 cm taken from the second layer of blue clay is not rich in diatoms and contains relatively many fragments and corroded valves. As in the sample at 265 cm the M-B-F ratio is 65-35-0 indicating no fully marine but brackish to strongly brackish conditions.

The assemblage of this sample is somewhat richer in complete valves of benthonic forms. This fact and the high clay content of the sediment point to sedimentation in a rather quiet high tidal area.

At 140 cm the second clay layer is still not rich in diatoms and the preservation of the valves resembles that of 165 cm. The M-B-F ratio is 55-40-5 which indicates again a decrease of the marine influence. The high value of the brackish forms (nearly all benthonic) shows the presence of stagnant, shallow brackish pools in a highly silted up area, which becomes more and more isolated from the marine influence.

Of the *Phragmites* peat overlaying the second clay layer, only the top could be examined at 119 cm. Here the peat is rich in diatoms and has a high content of fragments and corroded valves. The assemblage with a M-B-F ratio of 41-46-13 indicates peat formation under brackish conditions. This may mean that either the period of full isolation from the sea has passed already and the reed swamp is drowned again by a new marine transgression or there was no true period of full isolation at all. To answer this question more samples of this peat layer must be examined.

In the sample of 100 cm from the uppermost layer of blue clay a rather well preserved diatom assemblage occurs with a M-B-F ratio of 51-41-8. The assemblage indicates by the presence of a qualitatively rather rich brackish benthos, that on the peat layer again a high tidal brackish water clay has been deposited during a period of increasing marine influence.

In the sample at 85 cm from the same clay layer the diatom content decreases again and mostly fragments have been found. The M-B-F ratio of 67-30-3 shows a progressive increase of the marine influence. In the assemblage the brackish benthonic forms are not well represented but some of the truly marine forms occur in abundance. From this may be concluded that the sedimentation took place in a well exposed tidal area under at least strongly brackish conditions.

The sample at 60 cm from the transition clay-peat contains a badly preserved diatom assemblage with a M-B-F ratio of 54-42-4 indicating a new decline of the marine influence.

The assemblage is characterised by brackish benthonic species of highly silted up areas. The decrease of the marine influence gave rise to the forming of shallow brackish pools in which *Phragmites* could start growing.

The *Phragmites* peat of the uppermost sample of 55 cm is poor in diatoms and only fragments have been found. The M–B–F ratio of the assemblage is 44–31–25. The marine influence is still there but less important. The value of 25 for the indifferent fresh water species shows that the process of freshening advanced rapidly. The sedimentation took place under brackish to slightly brackish conditions.

Summary

In the Downholland profile three brackish to strongly brackish tidal clay layers are separated by *Phragmites* peat layers formed during periods of decreasing marine influence.

The peat between the two lowermost clay layers represents at least a period of nearly full isolation from the sea.

The peat layer at the basis of the uppermost clay layer partly (at the top) represents a period of less marine influence. More peat samples must be examined to trace whether there was a period of full or nearly full isolation during the growth of the lower part of the peat.

REFERENCES

AABY, B. (1975). Cykliske Klimavariationer de sidste 7500 år påvist ved undersogelser af hojmøser og marine transgressionsfaser. *Danm. geol. Unders. Årbog.* 1974 91–104.

——(1976). Cyclic variations in climate over the past 5500 years reflected in raised bogs *Nature, Lond.* **263** (5575): 281–4.

ADMIRALTY TIDE TABLES (1967, 1970). *European Waters, Including Mediterranean Sea.* Vol. I. Published by the Hydrographer of the Navy. xxviii + 357.

AGAR, R. (1954). Glacial and Post-Glacial Geology of Middlesborough and the Tees Estuary. *Proc. Yorks. Geol. Soc.* **29** (3:13): 237–53.

ANDREWS, J. T. and IVES, J. D. (1972). Late- and post-glacial events (10,000 BP) in the eastern Canadian Arctic with particular reference to the Cockburn moraines and break-up of the Laurentide ice sheet. *Acta Univ. Oul.* A **3**(1): 149–71.

——, KING, C. A. M., and STUIVER, M. (1973). Holocene sea level changes. Cumberland coast, north-west England: eustatic and glacio-isostatic movements. *Geologie Mijnb.* **52**(1), 1–12.

ASHMEAD, P. (1974). The caves and karst of the Morecambe Bay area. pp. 201–26. In WALTHAM, A.C., ed., *The Limestone and caves of north-west England.* Newton Abbot: David and Charles 477.

BARDEN, L. (1968). Primary and secondary consolidation of clay and peat. *Géotechnique* **18**: 1–24.

BARRON, J. (1938). *A History of the Ribble Navigation from Preston to the sea.* Preston: Preston Corporation at the Guardian Press, Fishergate. xv + 503.

BECK, J. (1954). The Church Brief for the Inundation of the Lancashire Coast in 1720. *Trans. Historic Society of Lancashire and Cheshire* **105**: 91–105.

BENZLER, J.-H. and GEYH, M. A. (1969). Versuch einer zeitlichen Gliederung von Dwog-Horizonten mit Hinweisen auf die Problematik der ^{14}C-datierung von Bodenproben. *Z. dt. geol. Ges.* **118**: 361–7.

BERGLUND, B. E. (1971). Littorina Transgressions in Blekinge, South Sweden: a preliminary survey. *Geol. För. Stockh. Förh.* **93** (3:546): 625–52.

BERRY, W. G. (1967). Salt marsh development in the Ribble estuary. pp. 121–35. In STEEL, R. W. and LAWTON, R., eds. (1967). *Liverpool Essays in Geography: a Jubilee Collection.* London: Longmans. ix + 600.

BINNEY, E. W. and TALBOT, J. H. (1843). *On the Petroleum found in the Downholland Moss, near Ormskirk.* Paper read at the Fifth Annual General Meeting of the Manchester Geological Society, 6 October 1843.

BISAT, W. S. (1952). Post-glacial peat and *Scrobicularia* clay, near Easington, Yorkshire. *Trans. Leeds Geol. Ass.* **6**: 210–14.

BLOOM, A. L. (1964). Peat Accumulation and Compaction in a Connecticut Coastal Marsh. *J. Sedim. Petrol.* 34(3): 599–603.

—— (1974). Sea-level history to be studied. *Geotimes* 1(1): 23–4.

BLUNDELL, N. (1720). The Great Diurnal of Nicholas Blundell Anno Domini 1720. pp. 1–32. In BAGLEY, J. J., ed. (1972). The Great Diurnal of Nicholas Blundell of Little Crosby, Lancashire. Vol. 3: 1720–1728. Transcribed and Annotated by F. Tyrer. *Record Society of Lancashire and Cheshire* **114**: xii + 289.

BOLTON, J. (1862). On a Deposit with Insect Leaves, etc. *Q. J. Geol. Soc. Lond.* **18**: 274–7.

BOTT, M. H. P. (1968). The Geological Structure of the Irish Sea Basin. pp. 93–115. In DONOVAN, D. T., ed. (1968). *Geology of Shelf Seas.* Edinburgh and London: Oliver and Boyd viii, + 160.

BRAND, G., HAGEMAN, B. P., JELGERSMA, S., and SINDOWSKI, K. H. (1965). Die lithostratigraphische Unterteilung des marinen Holozäns an der Nordseeküste. *Geol. Jb.* **82**: 365–84.

BRAY, J. R. (1970). Temporal Patterning of Post-Pleistocene Glaciation. *Nature, Lond.* **228** (5269): 353.

BRITISH STANDARDS INSTITUTION (1961). *Methods of Testing Soils for Civil Engineering Purposes.* British Standard 1377, 1961. 140.

BRYSON, R. A., WENDLAND, K. M., IVES, J. D., and ANDREWS, J. T. (1969). Radiocarbon Isochrones on the Distintegration of the Laurentide Ice Sheet. *Arctic and Alpine Research* 1(1): 1–14.

BUCKLEY, J. D., and WILLIS, E. H. (1969). Isotopes' Radiocarbon Measurements VII, *Radiocarbon.* 11 (1): 53–105.

CALLOW, W. J. and HASSALL, G. I. (1968). National Physical Laboratory Radiocarbon Measurements V. *Radiocarbon* 10: 115–18.

CHAPMAN, V. J. (1960). *Salt Marshes and Salt Deserts of the World.* London: Leonard Hill. xvi+392.

CHEESBROUGH, C. E., OLDFIELD, F., and PHILLIPS, A. W. (1969). *The Lytham St. Annes Foreshore.* A Research Project undertaken in the Department of Environmental Sciences of the University of Lancaster, on behalf of Lytham St. Annes Corporation. Final Report, June 1969. viii + 201.

CHURCHILL, D. M. (1965a). The Kitchen Midden Site at Westward Ho! Devon, England: ecology, age and relation to changes in land and sea level. With an appendix by J. J. Wymer. *Proc. Prehist. Soc.* 31 (5): 74–84.

—— (1965b). The displacement of deposits formed at sea-level, 6500 years ago in Southern Britain. *Quaternaria* 7: 239–47.

—— (1970). Post-Neolithic to Romano-British Sedimentation in the Southern Fenlands of Cambridgeshire and Norfolk. pp. 132–46. In PHILLIPS, C. W., ed. (1970). The Fenland in Roman Times: studies of a major area of peasant colonization with a Gazetteer covering all known sites and finds. *R.G.S. Res. Ser.* 5. xii + 359.

CLAPHAM, A. R. and GODWIN, H. (1948). Studies of the Post-Glacial History of British Vegetation. VIII. Swamping surfaces in the Peats of the Somerset Levels. IX. Prehistoric Trackways in the Somerset Levels. *Phil. Trans. R. Soc. B.* 233 (599): 233–73.

CLARKE, R. H. (1970). Quaternary sediments off south-east Devon. *Q. J. Geol. Soc. Lond.* 125: 277–318.

COLES, J. M., HIBBERT, F. A., and ORME, B. J. (1973). Prehistoric Roads and Tracks in Somerset: 3. The Sweet Track. *Proc. Prehist. Soc.* 39: 256–93.

COPE, F. W. (1939). Oil occurences in south-west Lancashire (with a Biological Report by Kathleen B. Blackburn). *Bull. Geol. Surv. G. B.* 2: 18–25.

CORLETT, J. (1972). The ecology of Morecambe Bay. I. Introduction. *J. appl. Ecol.* 9: 153–9.

DAMON, P. E., LONG, A., and GREY, D. C. (1970). Arizona radiocarbon dates for dendrochronologically dated samples. pp. 615–15. In OLSSON, I. U., ed. (1970). *Radiocarbon variations and absolute chronology.* Proc. 12th Nobel Symposium. Stockholm: Almqvist and Wiksell. 653.

DAMON, P. E., LONG, A., and WALLICK, E. I. (1973). Dendrochronological calibration of the Carbon-14 timescale. pp. A29–A43. In, RAFTER, T.A. and GRANT-TAYLOR, T. eds., (1973). *Proceedings of the Eighth International Conference on Radiocarbon Dating.* Wellington, New Zealand.

——, ——, ——, and FERGUSON, C. W. (1973). Dendrochronological calibration of the C^{14} time scale. pp. 75. *Abstracts 9th Congress of INQUA.* Christchurch, New Zealand.

DANSGAARD, W., JOHNSEN, S. J., MØLLER, S., and LANGWAY, C. C., Jr. (1969). One thousand centuries of climatic record from Cape Century on the Greenland ice sheet. *Science, N.Y.* 166 (3903): 377–81.

——, ——, REEH, N., GUNDESTRUP, N., CLAUSEN, H. B., and HAMMER, C. U. (1975). Climatic changes, Norsemen and modern man. *Nature, Lond.* 255 (5503): 24–8.

DELIBRIAS, G. and GUILLIER, M. T. (1971). The Sea-Level on the Atlantic Coast and the Channel for the Last 10,000 years by the ^{14}C Method. *Quaternaria* 14: 131–5.

DENTON, G. H., ARMSTRONG, R. L., and STUIVER, M. (1971). The Late Cenozoic

Glacial History of Antarctica. Ch. 10, pp. 267–306. In TUREKIAN, K. K. ed. (1971). *The Late Cenozoic Glacial Ages.* New Haven: Yale University Press. xii+606.

—— and KARLÉN, W. (1973). Holocene climatic variations—their pattern and possible cause. *Quaternary Res.* 3(2): 155–205.

DEVOY, R. J. N. (1977). 'Flandrian Sea-level Changes and Vegetational History of the Lower Thames Estuary.' Unpublished Ph.D. dissertation, University of Cambridge. 2 vols. ix+229.

DICKINSON, W. (1973). The development of the raised bog complex near Rusland in the Furness District of North Lancashire. *J. Ecol.* 61 (3): 871–86.

—— (1975) Recurrence surfaces in Rusland Moss, Cumbria (formerly North Lancashire). *J. Ecol.* 63 (3): 913–35.

DIGERFELDT, G. (1975). A standard profile for Littorina transgressions in western Skåne, South Sweden, *Boreas* 4: 125–42.

DONNER, J. J. (1970). Land/Sea Level Changes in Scotland. pp. 23–39. In WALKER, D. and WEST, R. 'G., eds. (1970). *Studies in Vegetational History of the British Isles: essays in honour of Harry Godwin.* Cambridge University Press. 266.

DUBOIS, G. (1924). Recherches sur les terrains quaternaires du Nord de la France. *Mém. Soc. géol. N.* 8 (1): 1–366.

DUPHORN, K., GRUBE, F., MEYER, K. D., STREIF, H., and VINKEN, R. (1973). Pleistocene and Holocene. pp. 222–50. In, BEHRE, K.-E. *et al.* (1973) State of Research on the Quaternary of the Federal Republic of Germany. *Eiszeitalter Gegenw.* 23/24: 219–370.

ENTE, P. J., ZAGWIJN, W. H., and MOOK, W. G. (1975). The Calais deposits in the vicinity of Wieringen and the geogenesis of northern north Holland. *Geologie Mijn.* 54 (1): 1–14.

EPSTEIN, S., SHARP, R. P., and GOW, A. J. (1970). Antarctic Ice Sheet: Stable Isotope Analyses of Byrd Station Cores and Interhemispheric Climatic Implications. *Science, N.Y.* 168 (3939): 1570–2.

ERDTMAN, G. (1926). Some micro-analyses of Moorlog from the Dogger Bank. *Essex Nat.* 21: 107–12.

—— (1928). Studies in the post-arctic history of the forests of North-west Europe. 'I. Investigations in the British Isles. *Geol. För. Stockh. Förh.* 50 (2:373): 123–92.

ERGIN, M., HARKNESS, D. D. and WALTON, A. (1972). Glasgow University Radiocarbon Measurements V. *Radiocarbon* 14 (2): 321–5.

ERONEN, M. (1974). The history of the Littorina Sea and associated Holocene events. *Comm. Physico-Mathematicae* 44 (4): 79–188.

EVANS, W. B. and ARTHURTON, R. S. (1973). North-west England. Ch. 6, pp. 28–36. In, MITCHELL, G. F. *et al.* (1973). A Correlation of Quaternary Deposits in the British Isles. *Geol. Soc. Lond.* Special Report No. 4. 99.

FAEGRI, K. and IVERSEN, J. (1964). *Textbook of pollen analysis.* New York: Hafner Publishing Co. 237.

FAIRBRIDGE, R. W. (1958). Dating the latest movements of the Quaternary Sea Level. *Trans. N.Y. Acad. Sci.* Series 2.20 (6): 471–82.

——(1961). Eustatic Changes in Sea-Level. pp. 99–187. In AHRENS, L. H., *et al.*, eds. (1961). *Physics and Chemistry of the Earth* 4. London: Pergamon Press. 317.

—— and HILLAIRE-MARCEL, C. (1977). An 8000 yr paleoclimatic record of the 'Double-Hale' 45 yr solar cycle. *Nature, Lond.* 268 (5619): 413–16.

FERGUSON, C. W., (1970). Dendrochronology of Bristlecone Pine, *P. aristata.* Establishment of a 7484 year chronology in the White Mountains of eastern-central California, U.S.A. pp. 303–12. In OLSSON, I. U., ed.(1970). *Radiocarbon variations and absolute chronology.* Proc. 12th Nobel Symposium. Stockholm. Almqvist and Wiksell. 653.

FISHWICK, H. (1907). The history of the parish of Lytham in the County of Lancaster. *Chetham Soc.* NS. 60: 1–118.

FORBES, C. L., JOYSEY, K. A. and WEST, R. G. (1958). On post-glacial Pelicans in Britain. *Geol. Mag.* 95 (2): 153–60.

FRENZEL, F. (1966). Climatic change in the Atlantic—Sub-Boreal transition in the northern hemisphere: botanical evidence. pp. 99-122. In SAWYER, J. S., ed. (1966). *World Climate 8000 to 0 B.C.* Proc. of the International Symposium held at Imperial College, London. 18-19 April 1966. London: Royal Meteorological Society. 229.

FULTON, R. J. (1971). Radiocarbon Geochronology of Southern British Columbia. *Geol. Surv. Can.* Paper 71-37. vi + 28.

GARRARD, R. A. (1977). The sediments of the south Irish Sea and Nymphe Bank area of the Celtic Sea. In, KIDSON, C. and TOOLEY, M. J., eds. 1977. *The Quaternary History of the Irish Sea.* Geological Journal Special Issue No. 7. Liverpool: Seel House Press. 345.

GAUNT, G. D. and TOOLEY, M. J. (1974). Evidence for Flandrian Sea-level changes in the Humber estuary and adjacent areas. *Bull. Inst. Geol. Sci.* **48**: 25-41.

GEYH, M. A. (1969). Versuch einer chronologischen Gliedening des marinen Holozäns an der Nordseekuste mit Hilfe der statistischen Auswertund von ^{14}C. Daten. *Z. dt. geol. Ges.* **118**: 351-60.

—— (1970). Middle and young Holocene Sea-Level Changes as global contemporary events *Geol. För. Stockh. Forh.* **93** (4): 679-90.

—— and STREIF, H. (1970). Studies on coastal movements and sea-level changes by means of the statistical evaluation of ^{14}C-data. pp. 599-611. *Proc. of the Symposium on Coastal Geodesy.* Munich.

GFELLER, D. and OESCHGER, H. (1963). Bern Radiocarbon Dates III. *Radiocarbon* **5**: 305-11.

GILLHAM, M. E. (1957). Coastal vegetation of Mull and Iona in Relation to Salinity and Soil Reaction. *J. Ecol.* **45** (3): 757-75.

GODWIN, H. (1940a). Studies of the Post-Glacial History of British Vegetation. III. Fenland Pollen Diagrams. IV. Post-Glacial Changes of Relative Land- and Sea-Level in the English Fenland. *Phil. Trans. R. Soc. B.* **230 (570)**: 239-303.

—— (1940b). Data for the study of Post-Glacial History. VI. A Boreal Transgression of the Sea in Swansea Bay. *New Phytol.* **39**: 308-21.

—— (1941). Studies of the Post-Glacial History of British Vegetation. VI. Correlations in the Somerset Levels. *New Phytol.* **40**: 108-32.

—— (1943). Coastal Peat Beds of the British Isles and North Sea. *J. Ecol.* **31** (2): 199-247.

—— (1948). Studies of the Post-Glacial History of British Vegetation. X. Correlation between Climate, Forest Composition, Prehistoric Agriculture and Peat Stratigraphy in Sub-Boreal and Sub-Atlantic peats of the Somerset Levels. *Phil. Trans. R. Soc., B.* **233 (600)**: 275-86.

—— (1954), Recurrence Surfaces. *Danm. geol. Unders.* II, **80**: 22-30.

—— (1955). Studies of the Post-Glacial History of British Vegetation. XIII. The Meare Pool Region of the Somerset Levels. *Phil. Trans. R. Soc. B.* **239** (662): 161-90.

—— (1956). *The History of the British Flora: A factual basis for Phytogeography.* Cambridge University Press. viii + 384.

—— (1959). Studies of the Post-Glacial History of British Vegetation. XIV. Late-Glacial Deposits at Moss Lake, Liverpool. *Phil. Trans. R. Soc., B.* **242 (689)**: 127-49.

—— (1960). Prehistoric wooden trackways of the Somerset Levels: their construction, age and relation to climatic change. *Proc. Prehist. Soc.* **26**: 1-36.

—— (1966). Introductory Address. pp. 3-14. In, SAWYER, J. S., ed. (1966). *World Climate from 8000 to 0 B.C.* Proc. of the International Symposium held at Imperial College, London. 18-19 April, 1966. London: Royal Meteorological Society. 229.

—— and CLIFFORD, M. H. (1938). Studies of the Post-Glacial History of British Vegetation. I. Origin and Stratigraphy of Fenland Deposits near Woodwalton, Hunts. II. Origin and Stratigraphy of deposits in Southern Fenland. *Phil. Trans. R. Soc. B.*, **229 (526)**: 323-406.

—— and GODWIN, M. E. (1933a). Pollen analysis of Fenland Peats at St. German's, near King's Lynn. *Geol. Mag.* **70 (826)**: 168-80.

——, —— (1933b). British Maglemose harpoon sites. *Antiquity* **7** (25): 36-48.

——, —— (1934). Pollen analysis of Peats at Scolt Head Island, Norfolk. pp. 64-76. In,

STEERS, J. A. ed. (1934). *Scolt Head Island: the story of its origin, the plant and animal life of the dunes and marshes.* Cambridge: Heffer and Son.

— —, — — and CLIFFORD, M. H. (1935). Controlling factors in the formation of Fen Deposits, as shown by Peat Investigations at Wood Fen, Ely. *J. Ecol.* **23(2)**: 509–35.

— — SUGGATE, R. P., and WILLIS, E. H. (1958). Radiocarbon dating of the Eustatic Rise in Ocean-Level. *Nature, Lond.* **181 (4622)**: 1518–19.

— — and SWITSUR, V. R. (1966). Cambridge University Natural Radiocarbon Measurements VII. *Radiocarbon* **8**: 390–400.

— — and WILLIS, E. H. (1959). Cambridge University Natural Radiocarbon Measurements I. *Am. J. Sci., Radiocarbon Supplement* **1**: 63–75.

— — — — (1960). Cambridge University Natural Radiocarbon Measurements II *Am. J. Sci., Radiocarbon Supplement* **2**: 62–72.

— — — — (1961). Cambridge University Natural Radiocarbon Measurements III. *Radiocarbon* **3**: 60–76.

— — — — (1962). Cambridge University Natural Radiocarbon Measurements V. *Radiocarbon* **4**: 116–37.

— — — — (1964). Cambridge University Natural Radiocarbon Measurements. VI. *Radiocarbon* **6**: 116–36.

GOLDTHWAITE, R. P. (1966). Evidence from Alaskan glaciers of major climatic changes. pp. 40–51. In, SAWYER, J. S., ed. (1966). *World Climate from 8000 to 0 B.C.* Proc. of an International Symposium held at Imperial College, London. 18–19 April, 1966. London: Royal Meteorological Society. 229.

GRAY, A. J. and BUNCE, R. G. H. (1972). The Ecology of Morecambe Bay. VI. Soils and Vegetation of the Salt Marshes: a multivariate approach. *J. appl. Ecol.* **9**: 221–34.

GRAY, J. (1965). Extraction techniques. Pt. III. Techniques in palynology. pp. 530–87. In KUMMEL, B. and RAUP, D. (1965). *Handbook of Paleontological Techniques.* San Francisco and London: W. H. Freeman and Co. xiii + 852.

GREENSMITH, J. T. and TUCKER, E. V. (1971a). The effects of Late Pleistocene and Holocene Sea-level changes in the vicinity of the River Crouch, east Essex. *Proc. Geol. Ass.* **82 (3)**: 301–22.

— — — — (1971b). Overconsolidation in some Fine-Grained Sediments; its Nature, Genesis and value in Interpreting the History of certain English Quaternary Deposits. *Geologie Mijnb.* **50(6)**: 743–8.

— — — — (1973). Holocene transgressions and regressions on the Essex coast, outer Thames estuary. *Geologie Mijnb.* **52 (4)**: 193–202.

GRESSWELL, R. K. (1953). *Sandy Shores in south Lancashire: The Geomorphology of south-west Lancashire.* Liverpool at the University Press. xii + 194.

— — (1957). Hillhouse Coastal Deposits in South Lancashire. *Lpool. Manchr. Geol. J.* **2**: 60–78.

— — (1958). The Post-glacial raised beach in Furness and Lyth, North Morecambe. *Trans. Inst. Br. Geog.* **25**: 79–103.

— — (1964). Western coast of Britain: Wales to the Lake District. pp. 247–63. In, STEERS, J. A., ed. (1964) *Field Studies in the British Isles.* London and Edinburgh: Nelson. xxiii + 528.

— — (1967) The Geomorphology of the Fylde, pp. 25–42. In STEEL, R. W. and LAWTON, R. eds. (1967). *Liverpool Essays in Geography: a Jubilee Collection.* London: Longmans. ix + 603.

HAGEMAN, B. P. (1969) Development of the Western Part of the Netherlands during the Holocene. *Geologie Mijnb.* **48 (4)**: 373–88.

HALL, B. R. (1954-5) 'Borehole Records from the Mosses of south-west Lancashire'. Soil Survey of England and Wales. MS. 65.

— — (1955). Lancashire. Sheets 74 (Southport) and 83 (Formby). *Report Soil Survey of England and Wales.* 1956. 7–9.

— — and FOLLAND, C. J. (1970). Soils of Lancashire. Soil Survey of Great Britain (England and Wales). *Bulletin* No. 5. Harpenden: Soil Survey of England and Wales. x + 173.

HAWKINS, A. B. (1971). The Late Weichselian and Flandrian Transgressions of south-west Britain. *Quaternaria* 14: 115-30.

HIBBERT, F. A., SWITSUR, V. R. and WEST, R. G. (1971). Radiocarbon dating of Flandrian pollen zones at Red Moss, Lancashire. *Proc. R. Soc. Lond., B.* 177: 161-76.

HODGSON, E. (1863). On a deposit containing Diatomaceae, Leaves, *etc.*, in the Iron Ore Mines, Near Ulverston *Proc. Geol. Soc. Lond.* 19 (1): 19-31.

HOGG, S. (1972). 'Post-Weichselian history of the north Northumberland coastal zone.' University of Durham, Department of Geography. Dissertation. 86.

HOWELL, F. T. (1965). 'Some aspects of the sub-Drift Surface of Parts of north-west England.' Unpublished Ph.D. Thesis. Manchester University.

—— (1973) The sub-drift surface of the Mersey and Weaver catchment and adjacent areas. *Geol. J.* 8: 285-96.

HUDDART, D. (1972). Late Devensian Glacial History. Ch. 2, pp. 3-47. In, HUDDART, D. and TOOLEY, M. J. eds. (1972). *The Cumberland Lowland Handbook.* Quaternary Research Association. v + 96.

—— and TOOLEY, M. J., eds. (1972). *The Cumberland Lowland Handbook.* Quaternary Research Association. v + 96.

——, —— and CARTER, P. A. (1977). The coasts of north-west England. In KIDSON, C. and TOOLEY, M. J., eds. (1977). *The Quaternary History of the Irish Sea.* Geological Journal Special Issue No. 7. Liverpool: Seel House Press. 345.

IVERSEN, J. (1937). Undersögelser over Litorinatrangressione i Danmark. *Medr. dansk. geol. Forenen.* 9: 223-32.

JAMES, R. (1636). *Iter Lancastrense.* Chetham Society 1845. Vol. XVII. 13 pp. Edited with notes and an introductory memoir by T. Corser. cxii + 13 + (17-85, Notes).

JARDINE, W. G. (1964). Post-glacial sea-levels in south-west Scotland. *Scot. Geog. Mag.* 80· 5-11.

—— (1971). Form and age of Late Quaternary shorelines and coastal deposits of south-west Scotland: critical data. *Quaternaria* 14: 103-13.

—— (1975). The determination of former sea-levels in areas of large tidal range. In, SUGGATE, R. P. and CRESSWELL, M. M., eds. (1975) *Quaternary Studies.* The Royal Society of New Zealand. *Bulletin* 13, 163-8.

JEFFERIES, R. L., WILLIS, S. J. and YEMM, E. W. (1968). The Late- and Post-Glacial History of the Gordano Valley, North Somerset. *New Phytol.* 67: 335-48.

JELGERSMA, S. (1961). Holocene Sea Level Changes in the Netherlands. *Meded. Geol. Sticht.* Serie C. VI. 7: 1-100.

—— (1966). Sea-level changes during the last 10,000 years. pp. 54-69. In, SAWYER, J. S., ed. (1966). *World Climate 8000 to 0 B.C.* Proc. of the International Symposium held at Imperial College, London, 18-19 April 1966. London: Royal Meteorological Society. 229.

—— de JONG, J., ZAGWIJN, W. H. and van REGTEREN ALTENA, J. F. (1970). The Coastal dunes of the western Netherlands; geology, vegetational history and archaeology. *Meded. Rijks. geol. Dienst.* N.S. 21: 93-167.

JENNINGS, J. N. (1952). The Origin of the Broads. *R. G. S. Res. Ser.* 2: 1-66.

JENSEN, H. (1972). Holocene sea-level and geoid-deformation. *Medr. dansk geol. Foren.* 21: 374-81.

JOHNSON, D. S. and YORK, H. H. (1915). *The relation of plants to tide levels. A study of factors affecting the distribution of marine plants.* Carnegie Institute, Washington. Publ. 206. 162.

JONES, E. W. (1959). *Quercus* L. Biological Flora of the British Isles. *J. Ecol.* 47: 169-222.

KARLÉN, W. and DENTON, G. H. (1976). Holocene glacial variations in Sarek National Park, northern Sweden. *Boreas* 5: 25-56.

KAYE, C. A. and BARGHOORN, E. S. (1964). Late Quaternary Sea-Level Change and Crustal Rise at Boston, Massachusetts, with Notes on the Autocompaction of Peat. *Bull. Geol. Soc. Am.* V. 75: 63-80.

KERR, D. B. (1967). 'The contribution of Arnside Moss to the Post Glacial History of Lowland Lonsdale, and some investigations into the Salt Marshes of North Morecambe Bay.' Unpublished B.A. Dissertation. Department of Environmental Studies, University of Lancaster.

KIDSON, C. (1971). The Quaternary History of the Coasts of south-west England, with Special Reference to the Bristol Channel Coast. Ch. 1, pp. 1-22. In, GREGORY, K. J. and RAVENHILL, W., eds. (1971) *Exeter Essays in Geography*. University of Exeter. xviii + 255.

— — (1977). The coasts of south-west England. In, KIDSON, C. and TOOLEY, M. J., eds. 1977. *The Quaternary History of the Irish Sea*. Geological Journal Special Issue No. 7. Liverpool: Seel House Press. 345

— — and HEYWORTH, A. (1973). The Flandrian Sea-Level rise in the Bristol Channel. *Proc. Ussher Soc.* 2 (6): 565-84.

KING-HELE, D. G. (1967) The shape of the earth. *Sc. Amer.* 217 (4): 67-76.

— — (1975). The Earth's gravitational field. In, *Geodynamics Today: a review of the earth's dynamic process*. London: The Royal Society. 197.

KOLP, O. (1976). Submarine Uferterassen der südlichen Cst-und Nord See als Marken des Holozänen Meeresanstiegs und der Überflutungsphasen der Ostsee. *Peterm. Geogr. Mitt.* 120: 1-23.

KOOIJMANS, L. P. L. (1974) *The Rhine/Meuse delta: Four studies on its prehistoric occupation and Holocene geology*. Leiden: E. J. Brill. xxiv + 421.

KROG, H. (1965) On the post-glacial development of the Great Belt. *Baltica* 2: 47-60.

— — and TAUBER, H. (1973). C^{14} chronology of late and post-glacial deposits in North Jutland. *Danm. Geol. Unders. Arbog.* 1973, 93-105.

KUKLA, J. (1969). The Cause of the Holocene Climate Change. *Geologie Mijnb.* 48 (3): 307-34.

LAMB, H. H. (1966). *The Changing Climate*. London: Methuen. 236.

LENNON, G. W. (1975). Coastal geodesy and the relative movements of land and sea levels. In, *Geodynamics Today: a review of the earth's dynamic processes*. London: The Royal Society. 197.

MACFADYEN, W. A. (1933). The Foraminifera of the Fenland Clays at St. German's, near King's Lynn. *Geol. Mag.* 70 (825): 182-91.

MACFARLANE, I. C. (1965). The Engineering Characteristics of Peat. Proc. 10th Muskeg Research Conference. 21-2 May 1964. National Research Council of Canada. Associate Committee on Soil and Snow Mechanics. *Tech. Mem.* 85.

McGRAIL, S. and SWITSUR R. (1975). Early British boats and their chronology. *Int. J. Nautical Archaeology and Underwater Exploration* 4 (2): 191-200.

MANLEY, G. (1965). Possible climatic agencies in the development of post-glacial habitats *Proc. R. Soc. Lond.* B. 161 (984) 363-75.

MARTIN, W. E. (1959). The Vegetation of Island Beach State Park, New Jersey. *Ecol. Mongr.* 29 (1): 1-46.

MAYR, F. (1964). Untersuchungen über Ausmass und Folgen der klime-und Gletscherschwankungen Seit dem Beginn der postglacien wärmezeit. *Z. Geomorph.* 9: 257-85.

MERCER, J. H. (1967). Glacier Resurgence at the Atlantic/Sub-Boreal Transition. *Q. J. R. Met. Soc.,* 93: 528-34.

— — (1978). West Antarctic ice sheet and CO_2 greenhouse effect: a threat of disaster. *Nature, Lond.* 271: 321-25.

MIKKELSEN, V. M. (1949). Praestø Fjord: the development of the post-glacial vegetation and a contribution to the history of the Baltic Sea. *Dansk. botanisk Arkiv.* 13 (5): 1-171.

MILLER, C. D. (1969). Chronology of Neoglacial Moraines in the Dome Peak area, North Cascade Range, Washington. *Arctic and Alpine Res.* 1 (1): 49-66.

MITCHELL, G. F. (1960). The Pleistocene History of the Irish Sea. *Advmt. Sci.* 17 (68): 313-25.

— — (1963). Morainic Ridges on the Floor of the Irish Sea. *Ir. Geog.* 4: 335-44.

—— (1972). The Pleistocene History of the Irish Sea: Second Approximation. *Scient. Proc. R. Dubl. Soc.* Ser. A. 4 (13): 181–99.

—— (1977). Raised beaches and sea-levels. Ch.13. pp. 169–86. In, SHOTTON, F. W., ed. (1977). *British Quaternary Studies: Recent Advances.* Clarendon Press: Oxford. xii+298.

MÖRNER, N-A. (1969a). The Late Quaternary History of the Kattegatt Sea and the Swedish West Coast. *Sver. geol. Unders. Afh.* Serie C. 640: 1–487.

—— (1969b). Eustatic and Climatic Changes during the last 15,000 years. *Geologie Mijnb.* 48 (4): 389–99.

—— (1972). Time, scale and ice accumulation during the last 125,000 years as indicated by Greenland O^{18} Curve. *Geol. Mag.* 109 (1): 17–24.

—— (1973). Eustatic changes during the last 300 years. *Paleogeogr. Palaeoclim. Palaeoecol.* 13 (1): 1–14.

—— (1976a). Eustatic changes during the last 8,000 years in view of radiocarbon calibration and new information from the Kattegatt region and other northwestern European areas. *Palaeogr. Palaeoclim. Palaeoecol.* 19: 63–85.

—— (1976b). Eustasy and geoid changes. *J. geol.* 84 (2): 123–51.

MORTON, G. H. (1887). Stanlow, Ince and Frodsham Marshes. *Proc. Lpool. Geol. Soc.* 5 : 349–51.

—— (1888). Further Notes on the Stanlow, Ince and Frodsham Marshes. *Proc. Lpool. Geol. Soc.* 6: 50–5.

—— (1891). *Geology of the Country around Liverpool, including the North of Flintshire.* London: George Philip and Son. vii + 319.

MORZADEC-KERFOURN, M.-T. (1974). Variations de la ligne de rivage Armoricaine au Quaternaire. *Mém. Soc. géol. minér. Bretagne.* 17: 1–208.

MOSELEY, F. (1972). A tectonic history of north-west England. *J. geol. Soc. Lond.* 128 (6): 561–98.

—— and WALKER, D. (1952). Some aspects of the Quaternary Period in North Lancashire. *Naturalist, Hull* April-June 1952: 41–54.

MURRAY, J. W. and HAWKINS, A. B. (1976). Sediment transport in the Severn estuary during the past 8000–9000 years. *J. geol. Soc. Lond.* 132: 385–98.

OELE, O. (1977). The Holocene of the western Netherlands. pp. 50–6. In, PAEPE, R. (1977) *Southern Shores of the North Sea.* Guidebook for Excursion C17. X INQUA Congress. Norwich: Geoabstracts. 63.

OLDFIELD, F.n.d. 'Appendix on the Results of Pollen Analytical Studies on the Foreshore of South Walney.' MS. 4 fos. + 2 Figs.

—— (1958). Technical data relating to the Pollen Analyses. Appendix I, pp. 98–102. In, GRESSWELL, R. K. (1958). The Post-Glacial Raised Beach in Furness and Lyth, North Morecambe Bay. *Trans. Inst. Br. Geog.* 25: 79–103.

—— (1960). Late Quaternary Changes in Climate, Vegetation and Sea-Level in Lowland Lonsdale. *Trans. Inst. Br. Geog.* 28: 99–117.

—— (1963). Pollen-analysis and Man's Role in the Ecological History of the South-East Lake District. *Geog. Annlr.* 45: 23–40.

—— (1965). Problems of Mid-Post-Glacial Pollen Zonation in Part of north-west England. *J. Ecol.* 53: 247–60.

—— and STATHAM, D. C. (1963). Pollen analytical data from Urswick Tarn and Ellerside Moss. *New Phytol.* 62: 53–66.

—— —— (1965). Stratigraphy and Pollen Analysis on Cockerham and Pilling Mosses, North Lancashire. *Mem. Proc. Manchr. lit. phil. Soc.* 107: 1–16.

PANTIN, H. (1977). Sediments of the north-eastern Irish Sea. In, KIDSON, C. and TOOLEY, M. J., eds. (1977). *The Quaternary History of the Irish Sea.* Geological Journal· Special Issue No. 7. Liverpool: Seel House Press. 345.

PEARSALL, W. H. (1918). The Aquatic and Marsh Vegetation of Esthwaite Water. V. The Marsh and Fen Vegetation of Esthwaite Water. *J. Ecol.* 6: 53–74.

PENNINGTON, W. (1970). Vegetation History in the north-west of England: A Regional Synthesis. pp. 41–79. In, WALKER, D. and WEST, R. G. eds. (1970). *Studies in the Vegetational History of the British Isles: essays in honour of Harry Godwin.* Cambridge University Press. 266.

— and LISHMAN, J. P. (1971). Iodine in Lake Sediments in Northern England and Scotland. *Biol. Rev.* 46: 279–313.

PICTON, J. A. (1849). The Changes of Sea-levels on the west coast of England during the Historic Period. Abstract. *Proc. lit. phil. Soc. Lpool.* 5: 113–15.

PORTER, S. C. and DENTON, G. H. (1967). Chronology of Neoglaciation in the North American Cordillera. *Am. J. Sci.* 265: 177–210.

POST, L. von (1933). A Gothiglacial transgression of the sea in south Sweden. *Geog. Annlr.* 15: 225–54.

— (1968). The Ancient sea Fjord of the Viskan Valley. Chapter X. Stages of Ancient Lake Veselången. *Geol. För. Stockh. Förh.* 90: 37–110.

RALPH, E. K. and MICHAEL, H. N. (1970). MASCA radiocarbon dates for *Sequoia* and bristlecone pine samples. pp. 619–23. In, OLSSON, I. U., ed. (1970). *Radiocarbon variations and absolute chronology.* Proc. 12th Nobel Symposium. Stockholm: Almqvist and Wiksell. 653.

RAMPTON, V. (1970). Neoglacial fluctuation of the Natozhat and Klutlan Glaciers, Yukon Territory, Canada. *Can. J. Earth Sci.* 7: 1236–63.

RANCE, C. E. de (1868). [Drift map of south-west Lancashire. 1:63,360] Geological Survey of England and Wales. Sheet No. 90SE. Drift surveyed 1868. Published 1873.

— (1869). The Geology of the Country between Liverpool and Southport. Explanation of Quarter Sheet 90SE. of the 1 inch Geological Survey Map of England and Wales. *Mem. Geol. Surv. U.K.* London: H.M.S.O. 15.

— (1872). Geology of the country around Southport, Lytham Southshore. Explanation of the Quarter Sheet 9°NE. *Mem. Geol. Surv. U.K.* 15.

— (1875). The Geology of the Country around Blackpool, Poulton and Fleetwood. Explanation of the Quarter Sheet 91SW. of the 1 inch Geological Survey Map of England and Wales. *Mem. Geol. Surv. U.K.* London: H.M.S.O. 14.

— (1877). The Superficial Geology of the country adjoining the coast of south-west Lancashire. *Mem. Geol. Surv. U.K.* London: H.M.S.O.

— (1878). Geology of the country around Preston, Blackburn and Burnley. Explanation of the Quarter Sheet 89NW of the 1 inch Geological Survey Map of England and Wales. *Mem. Geol. Surv. U.K.* London: H.M.S.O.

READE, T. M. (1871). The Geology and Physics of the Post-Glacial Period, as shown in Deposits and organic remains in Lancashire and Cheshire. *Proc. Lpool. Geol. Soc.* 2: 36–88.

— (1872). The post-glacial geology and physiography of west Lancashire and the Mersey estuary. *Geol. Mag.* 9 (93): 111–19.

— (1881). On a section of the Formby and Leasowe Marine Beds, and Superior Peat Bed, disclosed by cuttings for the outlet sewer at Hightown. *Proc. Lpool. Geol. Soc.* 4 (4): 269–77.

— (1902). Glacial and Post-Glacial Features of the Lower Valley of the River Lune and its estuary. *Proc. Lpool. Geol. Soc.* 9 (2): 163–93.

— (1908). Post-Glacial beds at Great Crosby as disclosed by the new outfall sewer. *Proc. Lpool. Geol. Soc.* 10 (4): 249–61.

REDFIELD, A. C. (1967). The Ontogeny of a Salt Marsh Estuary. pp. 108–14. In, LAUFF, G. H. ed. (1967). *Estuaries.* American Association for the Advancement of Science. Publication 83.

REID, C. (1913). *Submerged Forests.* Cambridge University Press. viii + 129.

ROSSITER, J. R. (1972). Sea-level observations and their secular variation. *Phil. Trans. R. Soc. Lond. A.* 272: 131–9.

SCHIEGL, W. E. (1972). Deuterium Content of Peat as a Palaeoclimatic Record. *Science. N. Y.* 175 (4021): 512–13.

SCHOLL, D. W. and STUIVER, M. (1967). Recent Submergence of southern Florida: A comparison with Adjacent Coasts and Other Eustatic Data. *Bull. Geol. Soc. Am.* **78**: 437-54.

SHAW, J.(1972). The Irish Sea glaciation of north Shropshire: some environmental reconstructions. *Field Studies* **3** (4): 603-31.

SHEPARD, F. P. (1963). Thirty-five thousand years of Sea Level. pp. 1-10. In, CLEMENTS, T. ed. (1963). *Essays in Marine Geology in Honour of K. O. Emery*. Los Angeles: University of Southern California Press.

—— and CURRAY, J. R. (1967). Carbon-14 Determination of Sea Level Changes in Stable Areas. pp. 283-91. In, SEARS, M. ed. (1967). *Progress in oceanography*. Vol. 4: *The Quaternary History of the Ocean Basins*. Oxford: Pergamon Press. xi + 344.

SHOTTON, F. W. (1967). The problems and contributions of methods of absolute dating within the Pleistocene period. *Q. J. Geol. Soc. Lond.* **122**: 357-83.

—— and WILLIAMS, R. E. G. (1971). Birmingham University Radiocarbon Dates V. *Radiocarbon* **13** (2): 141-56.

SIMPKINS, K. (1974). The late-glacial deposits at Glanllynnau, Caernarvonshire. *New Phytol.* **73**: 605-18.

SISSONS, J. B. (1966). Relative sea-level changes between 10,300 and 8,300 years BP in part of the Carse of Stirling. *Trans. Inst. Br. Geog.* **39**: 19-29.

—— (1967). *The Evolution of Scotland's Scenery*. Edinburgh: Oliver and Boyd. ix + 259.

—— (1972). Dislocation and non-uniform uplift of raised shorelines in the western part of the Forth Valley. *Trans. Inst. Br. Geog.* **55**: 145-59.

—— and BROOKS, C. L. (1971). Dating of Early Post-glacial Land and Sea Level Changes in the Western Forth Valley. *Nature Physical Science* **234** (50): 124-7.

SKEMPTON. A. W. (1970). The consolidation of clays by gravitational compaction. *Q. J. Geol. Soc. Lond.* **125**: 373-412.

SKERTCHLEY, S. B. J. (1877). The Geology of the Fenland. *Mem. Geol. Surv. U.K.* London: H.M.S.O. xvi + 335.

SMITH, A. G. (1958a). Post-Glacial Deposits in South Yorkshire and North Lincolnshire. *New Phytol.* **57**: 19-49.

—— (1958b). Two Lacustrine deposits in the south of the English Lake District. *New Phytol.* **57**: 363-86.

—— (1959). The Mires of South-Western Westmorland: Stratigraphy and Pollen Analysis. *New Phytol.* **58**: 105-27.

—— (1970). The stratigraphy of the Northern Fenland. pp. 147-64. In, PHILLIPS, C. W. ed. (1970). The Fenland in Roman Times: studies of a major area of peasant colonisation with a Gazetteer covering all known sites and finds. *R. G. S. Res. Ser.* **5.** xii + 359.

SPARKS, B. W. and WEST R. G. (1972). *The Ice Age in Britain*. London: Methuen, xvi + 302.

STARKEL, L. (1966). Post-glacial climate and the moulding of European relief. pp. 15-32. In, SAWYER, J. S., ed. (1966) *World Climate From 8000 to 0 B.C.* Proc. of the International Symposium held at Imperial College, London. 18-19 April 1966. London: Royal Meteorological Society. 229.

STEPHENS, N. and SYNGE, F. M. (1966). Pleistocene Shorelines. pp. 1-51. In, DURY, G. H. (1966). *Essays in Geomorphology*, London: Heineman. xi + 404 pp.

STILLMAN, C. J. (1968). The post-glacial change in sea-level in south-western Ireland: new evidence from fresh-water deposits on the floor of Bantry Bay. *Scient. Proc. R. Dubl. Soc.* Ser. A. **3**: 125-7.

STREIF, H. (1972). The results of stratigraphical and facial investigations in the coastal Holocene of Woltzeten/Ostfriesland, Germany. *Geol. För. Stockh. Förh.* **94** (2): 281-99.

SUESS, H. E. (1967). Bristlecone Pine Calibration of the Radiocarbon Time Scale from 4100 B.C. to 1500 B.C. In, I.A.E.A. (1967) *Radioactive Dating and Methods of Low Level Counting*. Proc. Symposium organised by I.A.E.A., Monaco, 2-10 March 1967.

—— (1970a). Bristlecone Pine calibration of the radiocarbon timescale, 5200 B.C. to the present.

pp. 303–12. In, OLSSON, I. U. ed. (1970). *Radiocarbon variations and Absolute Chronology*. Proc. 12th Nobel Symposium. Stockholm: Almqvist and Wiksell. 653.

—— (1970b). The three causes of secular C^{14} fluctuations, their amplitudes and time constants. pp. 595–604. In, OLSSON, I. U., ed. (1970) Op. cit.

SWINNERTON, H. H. (1931). The post-glacial deposits of the Lincolnshire coasts. *Q.J. Geol. Soc. Lond.* 87: 360–75.

TERS, M. (1973). Les variations du niveau marin depuis 10,000 ans, le long du littoral atlantique français. *Le Quaternaire: Géodynamique, Stratigraphie et Environment*. Centre National de la Recherche Scientifique. Comité National français de L'INQUA.

TERZAGHI, K. and PECK, R. B. (1967). *Soil Mechanics in Engineering Practice*. New York: John Wiley and Sons. xx + 729.

THOMAS, G. S. P. (1977). The Quaternary of the Isle of Man. pp. 155–78. In, KIDSON, C. and TOOLEY, M. J., eds. (1977). *The Quaternary History of the Irish Sea*. Geological Journal Special Issue No. 7. Liverpool: Seel House Press. 345.

TOOLEY, M. J. (1969). 'Sea-level Changes and the Development of Coastal Plant Communities during the Flandrian in Lancashire and adjacent areas'. Unpublished Ph.D. thesis, University of Lancaster. x + 160 pp.

—— (1970). The peat beds of the South-west Lancashire coast. *Nature in Lancashire* 1: 19–26.

—— (1971a). Evolution of the Fylde coast. pp. 1–7. In, WILSON, A. R. (1971). *Aspects of Fylde Geography*. Blackpool: Geographical Association.

—— (1971b). Changes in sea-level and the implications for coastal development. pp. 220–5. *Association of River Authorities Yearbook*. London.

—— (1973). Flandrian sea-level changes in north-west England and pan-North-west European correlations. pp. 373–4. *Abstracts*. International Union of Quaternary Research 9th Congress. Christchurch, New Zealand.

—— (1974). Sea-level changes during the last 9000 years in north-west England. *Geog. J.* 140: 18–42.

—— (1976a). The I.G.C.P. Project, Sea-level movements during the last 15000 years. *Quaternary Newsletter* 18: 11–12.

—— (1976b). Flandrian sea-level changes in west Lancashire and their implications for the 'Hillhouse coastline'. *Geol. J.* 11 (2): 37–52.

—— (1977a). *The Isle of Man, Lancashire Coast and Lake District*. Guidebook for Excursion A 4 (Ed. D. Q. Bowen), X INQUA Congress, 1977. Norwich: Geoabstracts. 60.

—— (1977b). The Quaternary of north-west England and the Isle of Man. pp. 5–7. In, TOOLEY, M. J. (1977a).

—— (1977c). Altmouth. p. 8. In, TOOLEY, M. J. (1977a).

—— (1977d). Drigg pp. 44–7. In, TOOLEY, M. J. (1977a).

—— (in the press). Holocene Sea-Level Changes: Problems of interpretation. *Geol. För. Stockh. förh.*

—— (in the press). The history of Hartlepool Bay. *Int. Jl. Nautical Archaeology and Underwater Exploration*.

—— and KEAR, B. (1977). Shirdley Hill Sand Formation. pp. 9–10, 11–12. In, TOOLEY, M. J. (1977a).

TRAVIS, C. B. (1926). The peat and forest bed of the south-west Lancashire coast. *Proc. Lpool. Geol. Soc.* 14: 263–77.

—— (1929). The peat and forest beds of Leasowe, Cheshire. *Proc. Lpool. Geol. Soc.* 15: 157–78.

TRAVIS, W. G. (1908). On Plant Remains in Peat in the Shirdley Hill Sand at Aintree, South Lancashire. *Trans. Lpool. Bot. Soc.* 1: 47–52.

—— (1922). On peaty beds in the Wallasey Sand-Hills. *Proc. Lpool. Geol. Soc.* 13 (3): 207–14.

TROELS-SMITH, J. (1955). Karakterisaring af Løse jordarter (Characterisation of Unconsolidated Sediments). *Danm. geol. Unders*. IV. Raekke. Bd. 3. 73 + 13 Tavler.

TURNER, C. and WEST, R. G. (1968). The subdivision and zonation of interglacial periods.

Eiszeitalter Gegenw. **19**: 93–101.

VALENTIN, H. (1953). Present Vertical Movements of the British Isles. *Geog. J.* **119 (3)**: 299–305.

VALENTINE, K. W. G. and DALRYMPLE, J. B. (1975). The identification, lateral variation, and chronology of two buried paleocatenas at Woodhall Spa and West Runton, England. *Quarternary Res.* 5: 551–90.

WALCOTT, R. I. (1972). Past sea levels, Eustasy and the Deformation of the Earth. *Quaternary Res.* **2** (1): 1–14.

WALKER, D. (1966). The Late Quaternary History of the Cumberland Lowland. *Phil. Trans. R. Soc.* B. **251 (770)**: 1–210.

WELIN, E., ENGSTRAD, L. and VACZY, S. (1972). Institute of Geological Sciences Radiocarbon Dates II. *Radiocarbon* **14** (1): 140–4.

WERFF, A. van der and HULS, H. (1958–66). *Diatomeeënflora van Nederland.* 8 parts. Published privately by A. van der Werff, Westzijde, 13a., De Hoef (U). The Netherlands.

WEST, I. M. (1972). The origin of the Supposed Beach at Porth Neigwl, North Wales. *Proc. Geol. Assoc.* **83** (2): 191–5.

WEST, R. G. (1970). Pollen zones in the Pleistocene of Great Britain and their Correlation. *New Phytol.* **69**: 1179–83.

—— (1972). Relative land-sea-level changes in south eastern England during the Pleistocene. *Phil. Trans. R. Soc. Lond.* A **272**: 87–98.

WHITTOW, J. B. (1960). Some comments on the raised beach platform of south-west Caernarvonshire and on an unrecorded raised beach at Porth Neigwl, North Wales. *Proc. Geol. Assoc.* **71** (1): 31–9.

—— (1965). The Interglacial and Post-Glacial strandlines of North Wales. pp. 94–117. In, WHITTOW, J. B. and WOOD, P. D. eds. (1965). *Essays in Geography for Austin Miller.* University of Reading.

—— (1971). Shoreline Evolution on the Eastern Coast of the Irish Sea. *Quaternaria* **12**: 185–95.

WILLIS, E. H. (1961). Marine Transgression Sequences in the English Fenlands *Ann. N. Y. Acad. Sc.* **95** (1): 368–76.

WISEMAN, J. D. H. (1966). Evidence for recent climatic changes in cores from the ocean bed. pp. 84–97. In, SAWYER, J. S. ed. (1966) *World Climate from 8000 to 0* B.C. Proc. of the International Symposium held at Imperial College, London, 18–19 April 1966. London: Royal Meteorological Society. 229.

WOLLIN, G., ERICSON, D. B. and EWING, M. (1971). Late Pleistocene Climates Recorded in Atlantic and Pacific Deep-Sea Sediments. Ch. 7, pp. 199–214. In, TUREKIAN, K. K., ed. (1971). *The Late Cenozoic Glacial Ages.* New Haven: Yale University Press. xii + 606.

WRAY, D. A. and COPE, F. W. (1948). Geology of Southport and Formby. One Inch Geological Sheets 74 and 83 NS. *Mem. Geol. Surv. U.K.* London: H.M.S.O. vi + 54.

WRIGHT, E. V. and CHURCHILL, D. M. (1965). The boats from North Ferriby, Yorkshire, England. *Proc. Prehist. Soc.* N.S. **31**: 1–24.

—— and WRIGHT, C. W. (1947). Prehistoric Boats from North Ferriby, East Yorkshire. *Proc. Prehist. Soc.* **7**: 114–38.

WRIGHT, W. B. (1914). *The Quaternary Ice Age.* London: Macmillan. xxiv + 464.

—— (1937). *The Quaternary Ice Age.* Second Edition. London: Macmillan. xxv + 478.

ZOLLER, H. (1960). Pollenalytische Untersuchungen zur vegetationsgeschichte der insurbischen Schweiz. *Denkschr. schweiz. naturf. Ges.* **83** (2): 45–156.

AUTHOR INDEX

PLACE INDEX

SUBJECT INDEX